THE BRITISH ECOLOGICAL SOCIETY

Ecological Issues Series

Land Management: The Hidden Costs

…erence entitled The hidden costs of modern land-…y the British Ecological Society and the European …alism as part of the 1997 British Association for …g in Leeds.

…ire (UK)

…ugh (UK)

…European Forum on Nature Conservation and …oralism is a non-profit making network which brings …ether ecologists, nature conservationists, farmers and …cy makers in order to promote an understanding of …high nature conservation value of certain farming …ems and to inform work on their maintenance.

British Ecological Society

Printed and bound in Great Britain at the University Press, Cambridge.

Contents

Key issues

- For decades, the apparent success of intensive land management and economic policies has involved hidden costs in the form of environmental, economic and social damage.
- Agricultural and conservation policies have tended to react to these problems symptomatically and piecemeal rather than deal with the underlying causes.
- This responsive approach has created an increasingly complex system of regulation and control that has been ineffective, costly to administer and often perverse in its effects.
- It is now recognised that in order to protect the environment and ensure long-term food security, policies need to establish systems of land management that exploit natural resources within locally sustainable limits.
- Traditional farming practices conform to these rules and have a proven record of sustainable production; moreover, they have created social, economic and biological diversity throughout Europe at a range of ecological scales.
- In order to prevent further damage to the environment and reduce risks to food security and thus ultimately human well being, there needs to be a policy shift from intensive production to more sustainable forms of land-management – possibly based on the use of locally adapted crops, stock and technologies.
- In time, this shift will integrate the interests of production and conservation and restore local distinctiveness in terms of cultures, economies, landscapes and biodiversity.

1 Introduction

Issues

Europe's countryside has undergone profound and damaging changes over recent decades, largely as a result of intensive, high input land management practices. Farmers had little say in the process - they either adapted to the new models of intensive, high-input farming or went out of business. In taking this route, however, they lost much of their self reliance, and have become dependent on subsidies, standardised production methods, and a declining choice of crops and animals. Their activities have also reduced the countryside to a biologically impoverished and uniform state, caused widespread pollution and are destroying one of farming's primary resources, the soil.

Although it is true that farmers have always attempted to maximise production, their efforts have traditionally been limited by the natural constraints of the land and the opportunities of the market. In recent years, however, Europe's land-use and economic policies have fixed market prices above free market values and provided farmers with other incentives to over-ride these natural constraints. This has polarised land-use, leading to a damaging pattern of overuse and neglect across most of Western Europe.

When production systems override natural constraints on a large-scale, as they have in the case of high-input land-uses, the environment becomes unstable and problems ensue. Maximising production, however, is not itself the problem. On the contrary, exploiting the productivity of the land within the natural limits imposed by climate, relief, soil and water, promotes biodiversity at a range of ecological scales. Indeed, the interaction between man and the land has produced a wealth of landscapes, habitats and farming systems. It has even created variation between and among domesticated breeds of plant and animal. This form of culturally-based biodiversity is unique, and because it is a product of human interaction with the land over thousands of years, it is, by definition, an important base-line for sustainable land-use management.

Intensive farming relies on high-inputs of fertilisers, pesticides and animal feed.

Given this simple relationship between sustainable production, high biodiversity and a healthy environment, it is remarkable that environmental protection has focused on the effects of intensification rather than its causes. Nature conservation, for instance, has been largely concerned with patching-up the casualties of high-input land-uses rather than dealing with the fundamental conflict between modern land-use and biodiversity. This has lead to a responsive, defensive system of protection that has compromising gaps, overlaps and contradictions.

Our failure to deal adequately with the envi-

Sustainable use means using landscapes, ecosystems, species and genes in a way that meets both present and future needs.

ronmental impact of land-use intensification, however, has not been without its lessons; indeed, the severity of the problems we face in terms of pollution, disease and system breakdown, have thrown these into high relief. For example, we now know that:
(a) there are natural limitations to the carrying capacity of the land, and that when we exceed these there are costs: we create a more fragile environment, and ultimately threaten our economic, social and physical well-being.
(b) by contrast with high-input systems, many of the surviving traditional land-use systems are not only efficient, sustainable means of production, but are inherently resilient and have produced a richness of biodiversity and beauty that would be impossible to create in any other way.
(c) land-use and economic policies are potent instruments of environmental change when they address causal issues, as they have in the case of land-use intensification; however, when they simply respond to the symptoms of a more general phenomenon, as they have through environmental legislation, they are largely ineffective.

These lessons not only (a) provide a compelling argument for drawing back from high-input land use practices; they also (b) provide well-tested models of alternative practice that optimise production and lead to varied and robust environment and (c) indicate the most effective way of making the necessary policy changes.

The importance of using our natural resources in a sustainable way is recognised by Article 6 of the internationally ratified Convention on Biological Diversity. This calls for the sustainable use of biodiversity to be integrated into plans, policies and strategies. Sustainability, it seems, is replacing productivity as the new policy imperative, and this is because it makes good sense. High-input systems may be productive in the short term, but many are inefficient, and damaging. Sustainable systems, by contrast, are a good long-term investment.

The measure of Europe's political commitment to sustainability can be seen in the ongoing reform of the Common Agricultural Policy (CAP). In pursuing the sustainable option, however, Europe's plans, policies and programmes need to recognise that locally-adapted low-input systems of production have been with us for generations, and are thus, by definition, sustainable (Info Box 1). The recently published (1998) EU Biodiversity Strategy already recognises the value of these systems for conserving wildlife; but their inherent sustainability also needs to be recognised—together with their ability to show us how to restore our damaged countryside and provide the necessary locally-adapted plants and animals.
The following chapters look at some of the

Info Box 1: Low-intensity farming systems

These systems of production use small amounts of fertilisers, pesticides and supplementary livestock feed. Their productivity is often low - relying on the natural fertility of the land and on the localised recycling of nutrients in the form of dung. The effectiveness of these systems often depends upon the use of locally adapted breeds of crop and animal, and on traditional skills and high inputs of manual labour. They have a long track record of success as productive systems and are generally associated with high levels of biodiversity.

'costs' of high-input land-use systems, and particularly the costs or impacts which undermine the viability of long-term food production (Section 2). They consider the inadequacy of our attempts both to protect our natural resources (Section 3) and to reverse the destructive impacts of high-input farming (Section 5). However, they also attempt to outline a more effective approach to both food production and conservation. This is discussed first in terms of the inherent sustainability of traditional systems of land-management and their association with high biological value (Section 4), and second by exploring the potential of economic reform for the reintroduction of sustainable methods into European land-use (Section 5).

The essential message of this publication is that, although robust and biologically diverse landscapes may be difficult to achieve by design, traditional land-use systems show that their development is a natural consequence of sustainable practice. In order to move towards this level of sustainability, farmers will need to be freed to farm in such a way that their activities are determined by the natural potential of the local environment and by factors that relate to what they produce. These include the market for food and other physical products, but also public demand for a healthy, attractive environment. Such has been the impact of intensification on traditional low-input systems, however, that it would be difficult to imagine how this transition could be made without the need for considerable investment in re-equipping suitable, locally-adapted management systems together with their rural infrastructure. One possibility would be to re-deploy European funds currently used to support intensive production. This would be an appropriate change in the use of public funds, and the level of support needed would tend to diminish as land-use sustainability increased.

2 Land-use intensification: the cost to the environment

An ecosystem is a dynamic complex of organisms and their environment working as a unit

Issues

Over the last half century, considerable environmental costs have been attributed to a European-wide system of public funding that favoured intensive farming and forestry over the more traditional land-use systems:

- Landscapes have been degraded by overuse or neglect, and in each case they have become more uniform

- Semi-natural ecosystems have either been lost or are inappropriately managed, and a wide range of plants and animals species have come under threat.

- Extensive areas of soil have been damaged by overuse, increasing erosion, flooding and drought, and posing a threat to the very basis of food production.

- Pollution has increased to the extent that water no longer meets health standards in many areas and needs special treatment, and production methods carry a threat of human disease.

- Traditional work patterns have disappeared and human communities dispersed, along with their locally developed skills, knowledge and cultures.

2.1 The drive for productivity

The impetus for funding land-use intensification emerged from Europe's recent history. World War II had major impacts on the civilian population, and shortages and food rationing continued well into peacetime. An understandable reaction to these privations was to become "self-sufficient" in terms of food and other raw

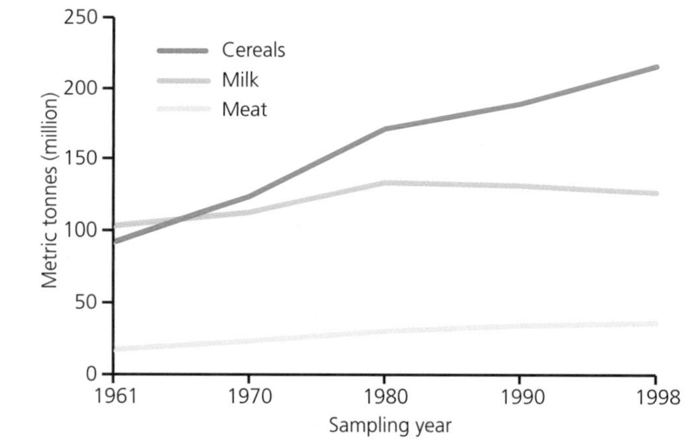

Figure 1. Change in total production of cereals, meat and milk. European Union (15). 1961-1998.UN FAOSTAT.

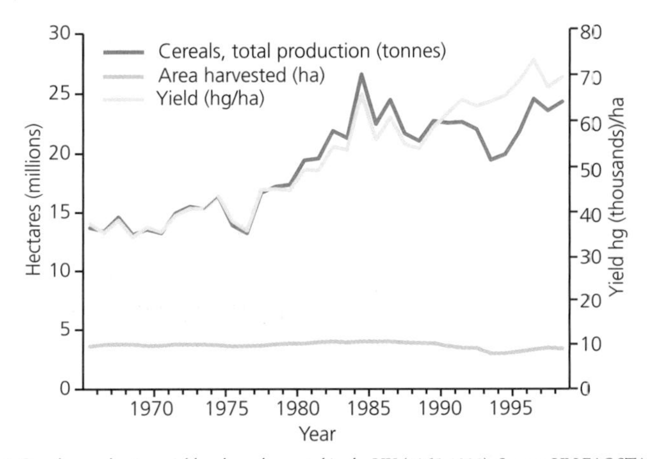

Figure 2. Cereals - production, yield and area harvested in the UK (1965-1998). Source: UN FAOSTAT.

materials such as timber, and this aim was encouraged by government funded support and incentives.

Throughout Europe, farming responded to these incentives by increasing the production (Figure 1) and yield (i.e. production per hectare or per animal) of a wide range of produce, particularly of cereals. In the UK, for example, cereal production almost doubled between 1965 and 1986, from 14 to 25 million tonnes by increasing yield per unit area (38,000 - 61,000 Hg/ha) (Figure 2). Even though production fell after peaking in 1986, yield continued to rise because of even more intensive production in some areas. This pattern of increase was common throughout much of the European Union (EU), culminating in the now familiar 'production mountains'. These may have reduced, but as late as 1993, Europe's cereal surplus was more than 30 million tonnes, and the beef intervention stores in 1998 were heading for 1 million tonnes.

Table 1. Fertilisers in the European Union (15 countries) and the UK: changes in consumption, imports, and cost of imports between 1963 and 1994. The UK figures are shown in (brackets) (U N FAOSTAT).

Total fertilisers	**1963**	**1994**	**change (%)**
Consumption (metric tonnes)	12,736,720 (1,495,200)	17,459,160 (2,219,800)	+37 +49
Imports (metric tonnes)	3,219,376 (615,600)	12,025,410 (1,461,500)	+274 +137
Imports as a % of consumption	25 (41)	68 (66)	
Cost of imports ($1000) Note: % change adjusted for inflation	346,168 (56,330)	4,481,722 (489,622)	+163 +78

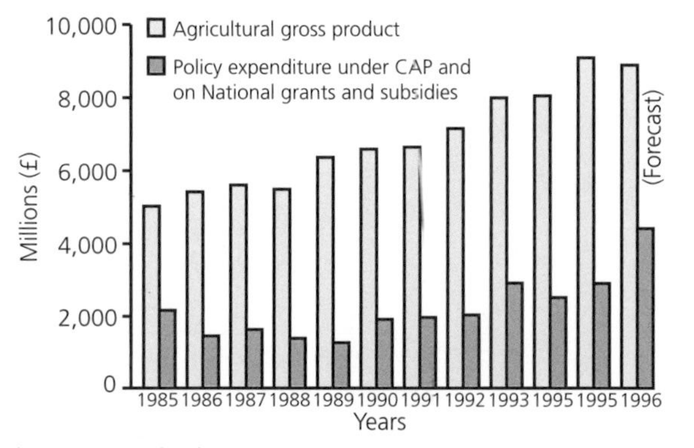

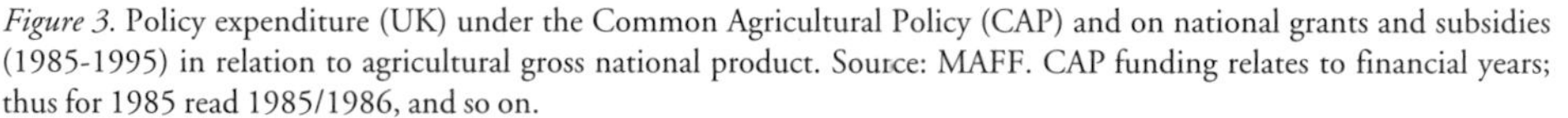
Figure 3. Policy expenditure (UK) under the Common Agricultural Policy (CAP) and on national grants and subsidies (1985-1995) in relation to agricultural gross national product. Source: MAFF. CAP funding relates to financial years; thus for 1985 read 1985/1986, and so on.

The post-war drive for self sufficiency was successful in terms of production levels, but it relied on heavy external inputs in the form of energy and agri-chemicals. These were needed to overcome the natural landscape constraints which had governed more traditional, lower-yielding, land-use systems. The irony is that this reliance on external inputs also rendered intensive farming systems vulnerable to strategic military and economic intervention - but for the import of material rather than food.

Between 1963 and 1994, for example, Europe not only increased its consumption of fertilisers by 37%, it more than doubled (25%-68%) the proportion provided by imports (Table 1).

High-input, intensive production methods have also involved heavy inputs of taxpayers' money in the form of expenditure under the Common Agricultural Policy (CAP). In the UK alone, this increased from just over £2bn in 1985 to £3bn in 1995, and is forecast to rise almost £4.5bn in 1996 (Figure 3) - almost half the agricultural gross product.

The land-use and economic policies that supported post-war increases in food production were born out of hardship and insecurity, and perhaps for this reason were rarely questioned at a strategic level. Even in the face of reasoned argument from environmentalists and farming associations, policy makers remained wedded to a rather narrow and contradictory approach to self sufficiency that depended on external inputs in the form of, energy, imported agri-chemicals, and public funding.

Even doubts about the true efficiency (Info Box 2) of intensive methods and mounting evidence about their environmentally damaging side effects,

Info Box 2: What is meant by efficiency?

The formal definition of efficiency is output per unit input, and traditionally it is measured in two forms of common currency: money (by economists) and energy (by ecologists and other scientists). However, simplistic attempts to reduce all values to one of these currencies have dangers. For example, the fallacy that intensive systems are necessarily more efficient than extensive ones results in part from including only the short-term costs of a farming enterprise regardless of its impact on the environment and the distorting effects of its financial structures (see Section 5). The relative efficiency of intensive and extensive systems will be discussed in Section 3.

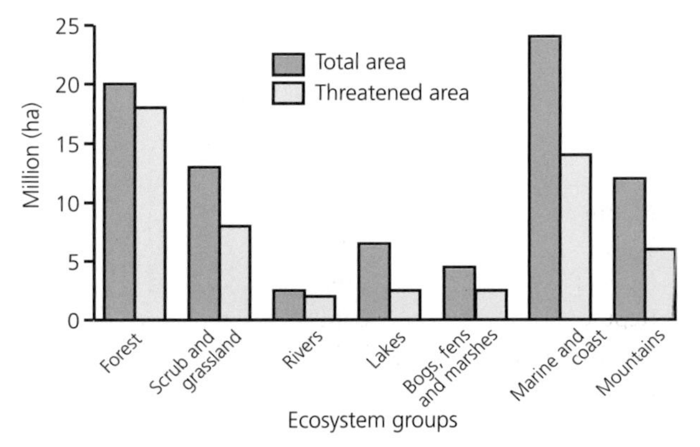

Figure 4. Representative sites of natural European ecosystem groups showing total area, and area where management problems and stresses threaten biodiversity; estimated from data in *Europe's Environment* 'Dobris'.

have failed to change the political view that they were viable means of production. This approach has persisted, despite official recognition that intensive practices had adverse environmental effects. In 1975, for instance, a UK government White Paper entitled 'Food from our own resources' called for an annual expansion in food production of 2.5%, stating that it would be possible to 'reconcile' this with environmental concerns. While the recognition of adverse affects was an important step forward in dealing with the problem, the reassurance was given without any justification, and has proved to have been over optimistic. Despite conservation's best efforts, the process of land-use intensification continues to have a profoundly damaging effect on the environment.

2.2 Habitats and wildlife

One of the most noticeable and publicised effects of land-use intensification, has been the loss of habitats and species. This has not been uniformly documented, but the European Environmental Almanac indicates that there have been widespread losses of pasture, coastal meadows (Portugal 80%), wetland (Spain 60%) and hedgerows (Ireland 16%). More significantly, the Dobris report shows that a large proportion of the remaining habitats have been compromised by a combination of environmental stress and management problems (Figure 4).

As might be expected, habitat changes, due to both intensification and abandonment, have affected a wide range of plants and animals. With birds it seems that some species, like the Carrion Crow (*Corvus corone*) and the Woodpigeon (*Columba palumbus*), have been able to withstand the effects of agricultural intensification, but many others like the Skylark (*Alauda arvensis*) have declined and are under threat. The dramatic decline in skylarks is probably typical of a wide range of farmland plants and animals throughout western Europe. Along with their habitats, many species are under a continuing threat - at least as great as elsewhere in the world (Table 2).

2.3 Overuse and neglect

One of the underlying features of the intensification process is a polarisation of land-use. This involves the over-exploitation of 'best and most versatile land' and the neglect of the more environmentally-limited areas. Across the whole of western Europe, this pattern of overuse and neglect has caused traditional, low-input farm-

Table 2. Threatened European Plants and animals set in a world context (Dobris 1995) estimates shown with *.

	World		Europe	
	Species total	threatened (%)	Species total	threatened (%)
Mammals	4327	16	250	42
Birds	9672	11	5620	15
Amphibians	4000	2	71	30
Reptiles	6550*	3	199	45
Fish (freshwater)	8,400*	4	227	52
Invertebrates	1,000,000*	?	200,000*	Unknown
Higher plants	250,000*	7	12,500*	21

Dobris is a comprehensive review of Europe's environment, covering topics as varied as pollution, population and biodiversity

ing systems to decline or be replaced by their intensive counterparts. The loss of these locally-adapted systems of production, involved a similar decline in their associated technologies, cultural systems, wildlife species, habitats, landscapes, and even in the range of domesticated species of plants and animals (Table 3).

Table 3 shows that land-use intensification is not only causing a general loss of habitats and wildlife, but it is simplifying the landscape of Europe, and causing a decline in biodiversity at many levels of the ecosystem. It also shows that intensification is causing pollution and soil erosion, which together with the loss of farm biodiversity, are undermining the economic basis of farming and thus posing a threat to food security.

2.4 Pollution

Land-use policies have concentrated animal production into small purpose-built units, increasing the pollution risks associated with handling slurry and silage liquor. They have also encouraged the use of agri-chemicals such as pesticides, herbicides and fertilisers, to the point where the timing, method and level of their use have caused them to leak into the wider environment as pollutants.

In the case of fertilisers, the European Environment Agency's Dobris Report estimates that the groundwater beneath more than 85% of Europe's farmland exceeds guideline levels for nitrogen concentration (25mg/l), with agricultural fertilisers being the main source of the problem. Surface water may also have similar problems. Indeed an Environmental Agency (UK) survey of 314 water bodies across England and Wales in 1994, found that over 50% had algal blooms caused by fertiliser run-off.

Pesticide and herbicide residues have also found their way into surface and ground water systems and, in the case of some of the older, pesticides such as Lindane and Dieldrin, are still being detected in areas where they have not been used for many years.

Although the cost of agricultural pollution is unlikely to be assessed comprehensively, it could be considerable. The cost of simply removing pesticides and nitrates from the water supplies in England and Wales, for instance, is known to have cost many millions of pounds. The reported figures (Table 4), however, are unlikely to reflect the full cost of installing the necessary water blending and nitrate removal equipment, which cost just one (UK) water company £80 million over four years.

Table 3. Ways in which the countryside of western Europe has been affected by the polarisation of land management practices (intensification and neglect) following the implementation of intensive land-use policies.

Intensification (most productive land)	Neglect (least productive land)
Social and economic impacts	
• increased level and intensity of production, but reduced efficiency of production (energy output/input) • replacement of locally adapted technologies by large-scale unified methods of production, and the replacement of locally distinctive produce by mass produced fare • replacement of established cultural infrastructures by simplified (ubiquitous) models geared to intensive production • importation of agri-chemicals and feed • deterioration in quality and the erosion of soil • reduction of genetic potential	• loss of traditional methods of production • decline of locally distinctive cultural infrastructures through the restructuring and abandonment of farms • loss of locally adapted indigenous technologies, and the decline of locally distinctive produce, through loss of processing and retailing facilities • reduction of genetic potential
Ecological impacts (landscape scale)	
• replacement of networks, corridors and mosaics by farming monocultures • sedentary patterns of production with a loss of transhumance • habitat fragmentation	• loss of mosaics networks and corridors through the successional development of scrub and forest (or by commercial forest monocultures) • decline of migratory forms of land management (transhumance, de-pasturing, folding) • habitat closure (which may amount to fragmentation - for example in the breaking-up of a patchwork landscape)
Ecological impacts (community scale)	
• loss of indigenous plant and animal communities (removal/nutrient enrichment) • loss of genetic variation among intensive breeds of crops and stock	• loss of indigenous plant and animal communities (succession/ nutrient enrichment) • loss of locally adapted, native breeds of crops and animals
Environmental impacts	
• air and water pollution - including nutrient enrichment • drought, flooding and siltation due to soil erosion. • disease due to ecologically unsound management practices (e.g. BSE) - worsened perhaps by high levels of genetic similarity	• threat from intensive forest monocultures • fires and subsequent erosion in Mediterranean areas

The amount of carbon stored in the top 30 cm of soil is double that in the atmosphere and three times that in the above-ground biomass (plant and animal matter)

Up to 3.2 million hectares of Europe, may be affected by some soil organic matter loss with localised reductions in yield of up to a third. The costs of fertiliser and manure needed to counterbalance this effect is not known.

Table 4. Cost of nitrate and pesticide removal at water treatment in England and Wales (millions of pounds 1997-98 prices).

	Nitrates	Pesticides
1994-95	20	134
1994-96	9	155
1996-97	5	117

Hansard: written answer 30/7/1998, Meacher.

2.5 Soil erosion

Soil is a complex, dynamic and open-structured medium consisting of mineral particles and organic matter in varying states of decomposition. It is home to a wide range of evolving and often mutually dependent organisms, some of which fix and store carbon and nitrogen and recycle nutrients. Its structural openness maximises the surfaces available for soil organisms, but also imparts a sponge-like quality which helps both drainage and moisture retention. This slows the movement of water through the environment, which along with its ability to fix atmospheric gases helps to stabilise the global environment. The primary value of soil, however, is its fundamental importance to a wide range of ecosystems - including those we use for crop production.

Farming systems affect soil conditions very differently. Traditional locally-adapted systems tend to maintain and improve soil condition by recycling nutrients (dung, plant and animal remains), often as part of a crop rotation. By contrast, intensive, high-input systems are often based on over-cultivated monocultures and depend on high levels of imported agrichemicals. Intensive management, therefore, tends to lower organic matter levels, releases nitrogen and carbon into the environment as pollutants and reduces the diversity of soil organisms.

Moreover, because organic matter helps to aggregates soil minerals into crumbs, its loss also reduces the soils porosity, structural stability and ultimately its ability to cope with intensive cultivation. Across much of Europe, this process has resulted in more fragile soil with an impaired ability both to drain and hold water - a situation that has increased the risk of seasonal drying, water-logging, surface run-off and erosion.

The Dobris report shows that up to 12% of Europe's total land area may be affected by water erosion and a further 4% by wind erosion - most of this (90%) occurring in the Mediterranean region. The full extent of the losses involved is not known, nor is their effect on crop production; however, a report prepared in 1996 for the Council for the Protection of Rural England and the World Wildlife Fund for Nature 'Sustainable Agriculture in the UK', indicates that the losses could be considerable. In England and Wales for instance, between 5-15% of the arable land may be losing 1-40 tonnes/ha of soil annually (against a soil formation rate of 1 tonne/ha/yr). The cost of this erosion has not yet been assessed for Europe, but Environment Canada (a government department) has estimated that the costs of on-farm water erosion in the potato belt of New Brunswick amounts to $10-12 million a year.

Even if there were no concern at the effect of intensive land-uses on biodiversity, or of the prospect of continuing to support with public funds the intensive production with imported fertilisers, there must be a fundamental objection to the idea of persisting with systems of production that not only pollute the environment, but also destroy soil, which is the very basis of agricultural production.

2.6 Loss of farm biodiversity

Intensive land-uses have not only simplified the ecology of Europe's 'natural' communities; they have also resulted in the loss of domesticated breeds of farm animal and varieties of cultivated crops. As with habitats and wildlife, there has been no comprehensive audit of these losses; however, the Food and Agriculture Organisation of the United Nations (FAO) estimates that over the last hundred years, the genetic diversity of the world's agricultural crops has declined by about three quarters, and over half of Europe's domestic animal breeds have become extinct.

A feature of intensive farming is that it has channelled most of its effort into a handful of productive breeds - the rest being largely neglected, except by special interest groups. This has reduced the background variation contributed by locally adapted domestic breeds and has placed the burden of food production on a limited range of 'elite' stock, which needs special husbandry, because it is not adapted to local circumstances. As a result, intensive farming has become dependent on the technologies and resources needed to support the productive stock, and the less productive indigenous farming systems have either been supplanted or are in decline. This loss of farm-based genetic diversity has both eroded the self-reliance of local farming communities and increased the vulnerability of agriculture to the challenges of disease and economic circumstances. It therefore constitutes a threat to 'food security' (Info Box 3), and as such, has important implications for economic and political decision making.

The effect of intensification on the biodiversity of farm crops is illustrated by winter wheat. Table 5 shows that 89 % of the weight of winter wheat seed certified from the 1995 UK harvest was produced by nine varieties, and almost half the harvest weight (49 %) produced by two varieties, 'Riband' and 'Brigadier'. It is also noticeable that the nine varieties have a high level of shared ancestry and were produced by four breeders. One breeder (PBI) produced four varieties which accounted for a quarter (24%) of the 1995 seed harvest.

As with farm crops, there has been a decline in the populations and range of farm animals. Commercial livestock farming has tended to use very few breeds of animal, ignoring the rest, except as novelties. Across Europe some breeds

Info Box 3: Food security and the genetic diversity of agricultural breeds and crops.

"Crop genetic diversity is not just a raw material for industrial agriculture; it is the key to food security and sustainable agriculture because it enables farmers to adapt crops suited to their own ecological needs and cultural traditions. Without this diversity, options for long-term sustainability and agricultural self-reliance are lost. The type of seed sown to a large extent determines the farmer's need for fertiliser, pesticides and irrigation. Communities that lose community-bred varieties and indigenous knowledge about them, risk losing control of their farming systems and becoming dependent on outside sources of seed and the inputs needed to grow and protect them. Without an agricultural system adapted to a community and its environment, self-reliance in agriculture is impossible." Food and Agriculture Organisation of the United Nations (1998). http:// www.fao.org/WAICENT/FAOINFO/ SUSTDEV/Epdirect/Epre0040.htm.

2

Table 5. Varieties of winter wheat accounting for the bulk of the UK 1995 seed harvest. Source: Ministry of Agriculture Fisheries and Food (UK) certification scheme (1996 Statistical Report). Details of parentage and breeder are also included: source National Institute of Agricultural Botany (NIAB, 1997).

Variety	Breeder	Parentage	Tonnes	
		Total seed weight for all varieties (62)	261,994	% Tot
Riband	PBI	Norman x (Maris Huntsman x TW161)	73,597	28
Brigadier	Zeneca	Squadron x Rendezvous	58,399	22
Hereward	PBI	Norman 'sib' x Disponent	21,772	8
Hussar	Zeneca	Squadron x Rendezvous	20,616	8
Consort	PBI	Riband 'sib' x Fresco x Riband	17,839	7
Hunter	PBI	Apostle x Haven	15,542	6
Siossons	Desprez	Jena x HN35	10,086	4
Rialto	PBI	Haven 'sib' x Fresco 'sib'	7,719	3
Buster	Nickerson	Brimstone x Parade	6,741	3
Total for the selected nine varieties			**232, 311**	**89%**
Total for the remaining fifty three varieties			**29, 684**	**11%**

Note: the weight of certified seed gives a good indication of what farmers sow - it is estimated, however, that farm-saved seed may be used for approximately 30% of the commercial area. Source NIAB, personal communication (Bradnam 1997).

have declined to the point where they now have dangerously small gene pools (Table 6). In the UK for instance it is recognised that more than thirty breeds of domestic animal have become extinct this century, with more than twenty breeds being classed as either 'critical' or 'endangered' (Table 7).

The impact of intensive farming on animal breeds is dramatically illustrated by dairy farming (Table 8). This shows that of the cows qualifying for inclusion in the 1995/1996 milk register for England and Wales, most were Holstein Friesian (95%). These have been selected for traits which succeed within the limited parameters of intensive commercial dairy farming, and are largely produced by a system of artificial insemination (AI). Using this technique, the ratio of bulls to cows can be very large. For example, Figure 5 shows that thirty six Holstein dairy bulls (four with the same sire) had produced 86,000 daughters by 1993.

Table 6. Breeds of domestic animals at risk in the European Union (15 countries, 1996). FAO/UNEP Environmental Statistics

	Cattle	Pigs	Sheep	Goats	Horses
critical	36	14	19	7	12
endangered	35	7	38	17	33

Definitions: Critical - breeding (females <100) (males <5)
Endangered - breeding (females 100-1000)(males >5 but < 20)

2

Table 7. United Kingdom breeds of cattle, sheep, pigs, ponies and horses that have become extinct during the 20th century, or are considered to be either 'critical' or 'endangered' (Rare Breeds Survival Trust, Stoneleigh).

Species	Extinct	Critical/ Endangered
Cattle	Alderney	Irish Moiled
	Glamorgan	Shetland
	Sheeted Somerset	Vaynol
	Pembroke	White Park
	Suffolk Dun	Gloucester
	Norfolk Red	
	Irish Dun	
	Caithness cattle	
	Castlemartin cattle	
Sheep	Rhiw	Castlemilk Moorit
	Ross Common	Galway
	Cannock	Norfolk Horn
	Morfe Common	Whitefaced
	Yel Shetland	Woodland
	St. Ronas Hill	Leicester Longwool
	Long Mynd	
	Gloucester sheep	
	Dunface sheep	
	Limestone sheep	
Pigs	Cumberland	British Lop
	Lincolnshire Curly Coated	Large Black
	Small White	Middle White
	Dorset Gold Tip	Tamworth
	Yorkshire Blue and White	Berkshire
	Ulster White	British saddleback
		Gloucestershire Old Spots
Horses/ponies	Cushendale	Cleveland Bay
	Long Mynd	Eriskay
	Manx	Suffolk
	Tiree	Exmoor Pony
	Galloway	
	Goonhilly	

Most of the daughters (82,000) were produced by five bulls, and one bull, Paltzer Sexation Bert, produced over half these (54,000). Bert's output is by no means exceptional in the AI industry; however, it is much larger than what might be expected under 'natural' herd conditions, and it represents a considerable focusing of reproductive effort. Although there is not a great deal of published data on the effective gene pool of the cattle breeds that underpin intensive farming, it is possible that it may be as narrow as some of our more endangered domestic breeds.

With both crops and animals, the breeds selected for intensive farming are generally limited

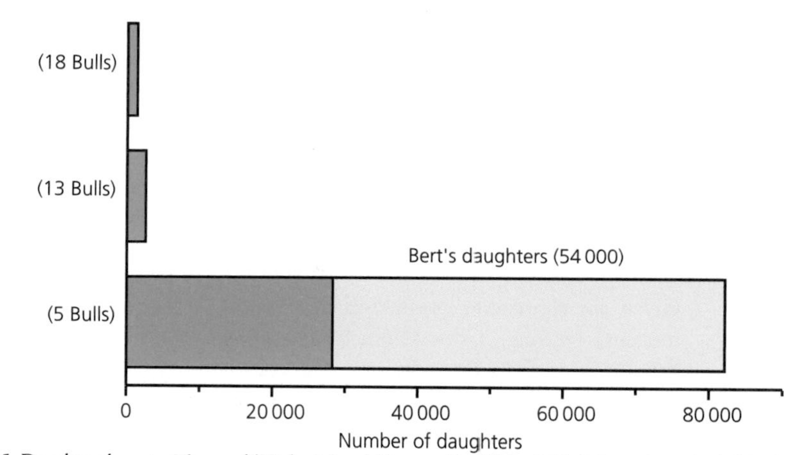

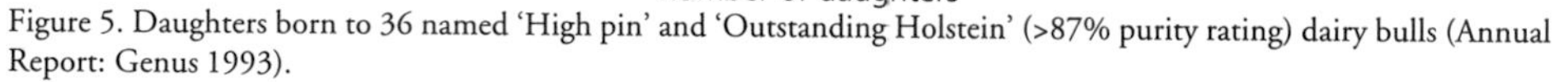

Figure 5. Daughters born to 36 named 'High pin' and 'Outstanding Holstein' (>87% purity rating) dairy bulls (Annual Report: Genus 1993).

to those that can respond rapidly to high-input systems of productions. They are thus adapted to the simplicity of a few key inputs rather than the subtleties of the 'natural' environment. In the case of stock, many highly bred animals now lack the ability to develop adequately on a diet of naturally-occurring low-grade fodder because of their small rumen size. Even on a

Table 8. Breeds of dairy cow that account for the bulk of lactations (cows and heifers) in England and Wales.

Breed	**Average milk yield (kg)**	**Total lactations (Cows and Heifers)**	
		Number	**Percentage**
Holstein Friesian	6638	963,559	94.5
Guernsey	4703	10304	1.0
Ayrshire	5822	13105	1.3
Jersey	4491	18719	1.8
Shorthorn	5587	4378	0.4
Island Jersey	4324	3985	0.4
Island Guernsey	5125	1674	0.2
Meuse-Rhine-Issel	5217	1003	0.1
Brown Swiss	6010	946	0.1
Simental	5205	368	0.04
Red Poll/Red Dane	4931	214	0.02
Montbeliarde	5824	309	0.03
Devon/South Devon	4726	44	0.004
Total lactations qualifying for production report		**1019187**	**100**

The numbers indicate dairy cows that qualified for inclusion in the 1995/96 annual production report of the national milk register. This amounts to sixty percent of the cows in England and Wales.

diet of high grade fodder, they may fail to thrive without high energy supplements - much of which are imported from outside Europe. This reinforces the view that Europe's farming communities are losing control of their farming systems at a local level and are becoming dependent upon external, often imported, inputs.

In 1998, a House of Lords (UK) select committee recommended phasing out livestock growth promoters that are related to ant-imicrobial agents for use in man.

Given the short-term commercialism behind intensive farming, it should not be surprising that antibiotics are used to promote animal growth and prevent disease due to overcrowding and genetic similarity. Nor should it be surprising that animals have been fed with the processed remains of closely related species. However, from an ecological perspective, it should be less surprising that the long term implications of these practices, are beginning to reveal themselves in the form of antibiotic resistance and BSE (Bovine Spongiform Encephalopathy).

Between 1996 and 1998, BSE related support to the beef industry cost £2.5 bn. The treatment of human blood supplies to reduce the 'remote' risk of transmitting the new varient Creutzfeldt-Jacob disease (human form of BSE) will cost the UK an estimated £80 m a year.

In the case of BSE, the effects of intensification have been both expensive and tragic. Not only has BSE seriously damaged the intensive beef industry, the restrictions imposed to control the disease have even affected farm incomes and wildlife habitats in the more environmentally friendly farming systems where the disease has not occurred. New variant BSE is also having an impact on medical services, and may yet have serious long-term implications for human health. Whatever the cost of BSE to agriculture and the rural economy may be, it could be dwarfed by the need to deal with any long-term effects it poses for human health.

2.7 Impact of land-use intensification on forest biodiversity

The pattern established by the industrialisation of agricultural intensification has been repeated by industrial forestry. This has used a variety of financial support mechanisms, including tax incentives to plant considerable areas across Europe. Much of this new planting, whether in existing wooded landscapes or new plantations, made little pretence at emulating nature in terms of stand structure, species composition, or for that matter in the pattern or scale of regeneration. The emphasis was on creating large-scale plantations of single-species, even-age conifers - often raised from imported seed. Figure 6 shows the area of woodland recorded

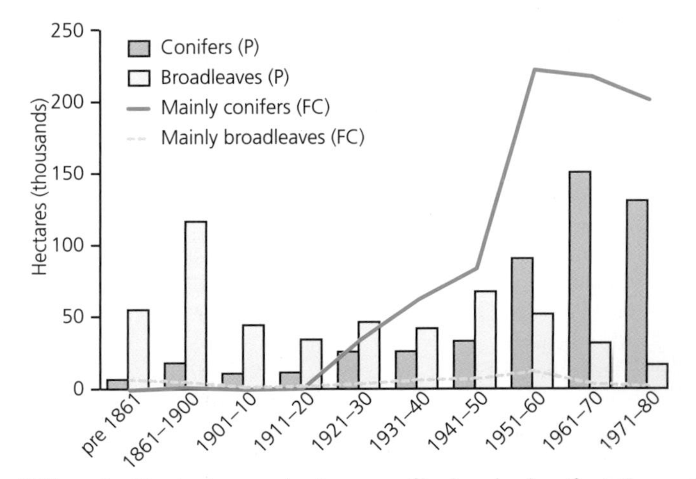

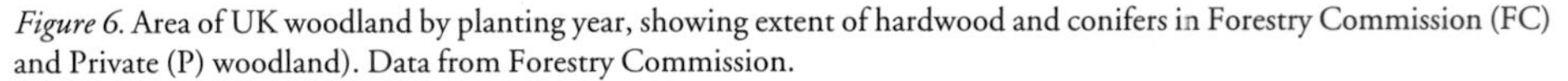
Figure 6. Area of UK woodland by planting year, showing extent of hardwood and conifers in Forestry Commission (FC) and Private (P) woodland). Data from Forestry Commission.

2

Industrial forestry uses large-scale, uniformly-structured plantations of fast growing trees harvested over short time-scales for industrial products, such as pulp-wood, saw-logs and poles.

Provenance means place of origin. Since traits within a species can differ from region to region, it is an important aspect of biodiversity.

Trees may yield very little seed for many years and then have a 'mast' year, when a given species produces an abundance of seed over a wide geographic area.

in the 1979-1982 woodland census (UK) by planting year class and landowner (Forestry Commission or Private). This indicates a high level of conifer planting from about 1941 for both types of landowner. However, while the UK Forestry Commission used conifers almost exclusively, private owners continued to plant a substantial area of broadleaves until the late 1950s. The example set by the government agency in the use of conifers, fits well with a state, policy-driven model of intensification.

Even where commercial forestry has used native trees, the reproductive material was selected from 'elite' sources - sometimes with a remote provenance. The selection of 'productive' oak and beech seed is governed by law in the UK under the Forestry Reproductive Material Act. The vagaries of masting, however, meant that it has often been necessary to collect seed from unregistered woods. Together with the inherent variability of the seed, and the natural regeneration of local ecotypes, this has helped to confound the worst excesses of intensification in the cultivation of native tree species.

The search for 'elite' native stock, however, has involved a considerable gene-flow across countries and continents, and this has fuelled a growing concern about the ecological importance of provenance in the selection of forest reproductive material. The European Forests Genetic Resources Programme for instance, considers that as well as being selected for productivity, commercial stock should be 'source identified and used in a way which was consistent with the genetic and ecological characteristics of its provenance'. Similarly, the biodiversity strategy of the European Commission also argues for the appropriate use of native species with a local provenance.

The relative economics of forestry and farming ensured that much of the new commercial planting took place on land that was suitable only for extensive, low-input farming - often within complexes of 'poorer quality' pasture, woodland and heathland. Table 9 gives some idea of the likely extent of the replacement that has occurred. This shows that between 1963 and 1994, forest cover increased in Europe as a whole by 11% and in the UK by 36%, while all other rural land-uses decreased by between 10% and 19%. This increase in forest cover was

Table 9. Change in rural land-use in the European Union (15 countries) and the UK between 1963 and 1994. The UK figures are shown in (brackets). (UN FAOSTAT).

Total land area (1000ha)	**1963**	**1994**	**change (%)**
Agricultural Area	163,545	143,764	-12
	(19,747)	(17,046)	-14
Arable	87,227	76,773	-12
	(7,289)	(5,902)	-19
Permanent Pasture	65,008	56,225	-14
	(12,348)	(11,097)	-10
Woodland & plantations	102,122	113,251	+11
	(1,758)	(2,390)	+36

not achieved at the expense of intensively farmed land, but rather was part of a process which replaced biologically valuable and potentially sustainable land-use systems with large-scale single-aged monocultures of exotic trees.

These largely state-supported forestry initiatives have not only replaced existing habitats but have disrupted ancient ecosystems at a landscape scale. For example, the enclosed commercial planting in the New Forest, Hampshire (UK) has 'fixed' a substantial proportion (>20%) of what was essentially a large-scale, shifting mosaic of pasture and woodland. These conifer plantations not only displace and fragment the native pasture woodland complex, but also impede the movement of forest animals - a fundamentally important aspect of any grazing ecosystem.

Currently (1994-1997) there are plans to afforest 1,000,000 ha of farmland throughout Europe and 'improve' farm forestry on a further 700,000 ha of established woodland. Between 50-75% of the costs involved in this programme is being supported under regulation (EC) 2080/92, as part of the European Union Forest Strategy. While this initiative is potentially an important aspect of the rural development process, it could in fact provide a considerable incentive to afforest extensive low-input farm land, and thus further undermine the viability of long-established and sustainable land-use systems. For this reason, it is hoped that Europe's reforestation programme will not repeat the pattern of earlier forestry initiatives, but will instead, respect the integrity of other important ecosystems, and most importantly the land management processes that produce them. Indeed, a priority for any future integrated rural development initiatives should be to start the integration process with the two principal land-uses, namely agriculture and forestry.

2.8 Landscape effects of intensive forestry and farming

Intensive production in both farming and forestry has tended to produce a less complex, more uniform landscape adapted to a few simple inputs. This new land-use has simplified the landscape in two main ways.

On the most productive (and generally more versatile) land, intensive production has destroyed or fragmented the established infrastructure of woods and hedgerows and increased drainage, or in the case of arid areas, provided large-scale irrigation. It has also confined the use of crop varieties and breeds of domestic animal to a relatively few highly productive types.

On the less productive land, traditional systems of low intensity production have remained, but these are declining and have become vulnerable to industrial forestry and other simplifying forces, such as neglect. In some areas, the sale of the less productive agricultural land for plantations has funded the intensification of valley land, leading to the break-down of the traditional systems of production and the further simplification of the environment at a landscape scale.

Over the whole of Europe, the combined effect of intensification and neglect is eroding systems of production that have adapted to local conditions over hundreds, and sometimes thousands, of years. These formerly common systems emerged from a long interaction between humans and the environment. Because of this, they are inherently sustainable and have a distinctive biology; however, their survival depends on the continuance of traditional low-input land management practices.

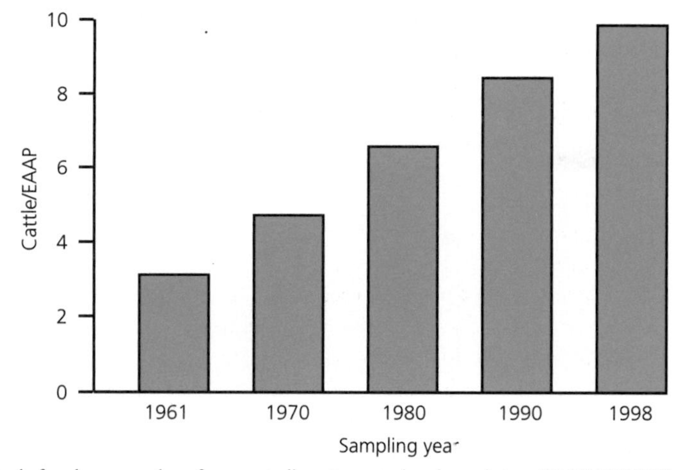

Figure 7. Head of cattle per member of economically active agricultural population. (UN FAOSTAT).

2.9 Cultural effects

The polarisation of land use resulting from the intensification of land management has not only had a profound effect on the fabric of the countryside and its dependent species but also on its cultural systems and indigenous technologies. A consequence of the industrialisation of farming and forestry is a greatly reduced need for human input. This effect is generally well understood, and is indicated by an increase in cattle per head of the farming population (Figure 7). What is less well recognised, however, is the scale of change in recent years. Table 10 gives some idea of the changes that took place in the UK in the 10 years from 1986. This shows losses at all levels of the farm work force (-12%), with by far the greatest decline occurring among the regular full-time workforce (-32%). In the most intensively farmed areas, depopulation has been even more dramatic.

In the marginalised, less-fertile areas of Europe, intensification has sometimes resulted in complex land-use changes. Some areas have been intensified, others have been transferred to another use, such as forestry, or even abandoned.

Table 10. Number (1000s) of farmers and workers (UK) (1986-1995).

	1986	1996	% change
Total farm labour force	682.5	603.2	-12
Total farmers (doing farm work)	290.6	280.9	-3
Spouses, partners, directors	77.1	74.6	-3
Salaried managers	8.3	7.8	-6
Regular whole-time workers	148.9	101.6	-32
Regular whole-time workers	62.0	56.7	-9
Seasonal or casual workers	95.6	81.7	-15

Source MAFF (UK).

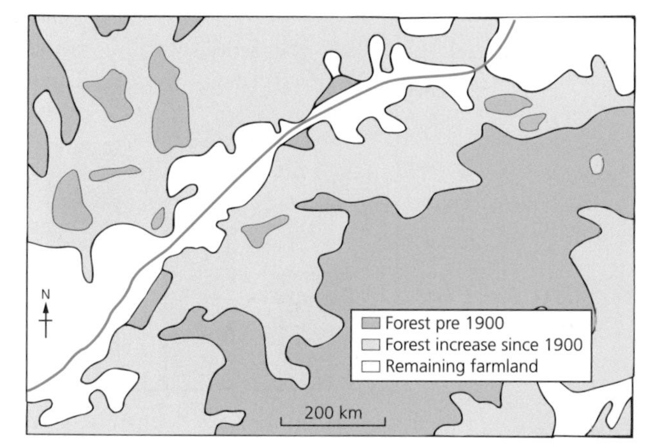

Figure 8. Prechtal Valley in the Central Black Forest, Germany - replacement of farmland (mainly grassland) by plantations (mainly Spruce) since 1900. Source: Rainer Luick (The common agricultural policy and environmental practices).

Transhumance involves the seasonal movement of stock between different types of grazing. It is probably a human adaptation of wild migration patterns

In the Black Forest, this process has halved the number of full-time and part-time farms have since 1974, allowing trees, scrub and plantations to encroach on to biologically rich pasture (Figures 8, 9). It has also eroded the local skills base. In the Freiberg region, for example there has been a fall in the number of agricultural trainees since the second world war, and from 1970, the number fell below the minimum (250/year) needed to secure the future of the 8000 full-time farms in the region (Figure 10).

Much of Europe's low intensity farming systems have been affected in a similar way to that of the Black Forest, resulting in the decline of traditional systems of land management and a break-up of established communities. In Greece, Italy and Spain, intensification has caused a shift from transhumance to a more sedentary form of livestock management, usually encouraged by subsidies and the availability of feed concentrates. In Mavrolithariou

Figure 9. Prechtal Valley, showing forest encroachment due to the withdrawal of low-input agriculture followed by farm closures and EU-supported afforestation. Only a few valleys in the region now remain open. Rainer Luick.

The precautionary approach resists change that may compromise the welfare of future generations

(Greece), for instance, the seasonal movement of livestock fell by a quarter between 1980 and 1991. There was also a 16% decline in the number of herders, and the average age of farmers rose from 57 to 63 years. The impact of these changes on cultural systems and indigenous technologies are profound, as indeed they are on the landscape.

2.10 Concluding thoughts

Although the effects of intensive practices are now being recognised in terms of their crude effects on the habitats, landscapes and cultural systems of Europe, it would be a mistake to imagine that intensification will soon be a thing of the past. The short-term imperatives of commerce and politics remain, and both farming and forestry retain the capacity to intensify through the medium of biotechnology - perhaps to a greater degree than ever.

In the past, breeding barriers tended to restrict the transfer of genes between distantly related groups; this is no longer the case - evolutionary determined gaps in the ecosystem can now be bridged relatively easily to produce 'genetically modified crops' (GMC). While it will remain easy to provide clear-cut, short-term financial arguments to promote GMC technologies, especially since it can be argued that they may help to reduce farming inputs, the ecological arguments for restraint will be difficult to articulate. This is simply because natural systems are complex and do not always react quickly to unfavourable disturbance. For this reason, a precautionary approach will need to be taken to new releases. Measures intended to foster this approach are to be embodied in a European protocol on biosafety. This is currently (1998) being finalised by the European Commission.

The damaging effects of not seeing the big picture in the way that land management and the environment interact are evident throughout Western Europe. Despite this, there is still no ecological model that links causal processes with environmental effects. The situation not only makes it difficult to assess the likely impact of land-use policies and developments such as genetically modified crops; it compromises our approach to environmental protection. It does this by leaving us to deal with the adverse consequences of land-management change, piecemeal and symptomatically. The difficulties with this approach to conservation are considered in Section 3, before reviewing alternative solutions in later sections.

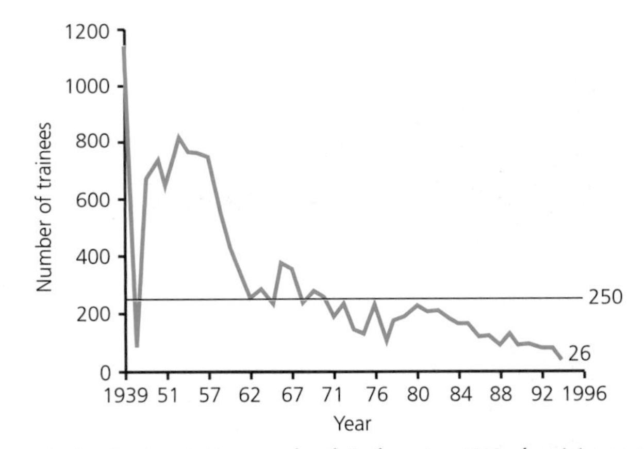

Figure 10. Decline in agricultural trainees in Regierungsbezitk Freiberg since 1939 - the minimum annual intake needed to secure the future of the region's 8000 full-time farms is 250. Source: Rainer Luick (The common agricultural policy and environmental practices).

2

Summary

Intensive practices may be productive in the short-term, but they are not necessarily efficient - at least not when their side-effects are taken into account, along with the cost of the high inputs of energy, feed and fertilisers.

Moreover, their impact on the landscape and its related cultural systems and biodiversity throws into question their long-term sustainability; indeed, their destructive impact on the soil, the very source of its productivity, means that intensive production is not only degrading the environment but is self-destructive.

Current methods of land management need to be replaced by sustainable and socially acceptable methods of food production which embrace the needs of conservation and environmental protection, and which also reflect the rich diversity of Europe's cultural landscapes. This means: (a) farming within the natural limitations of the local environment; (b) limiting external inputs and consequently accepting lower production levels; and (c) taking precautionary approach to the application of new intensive practices.

3 Biological conservation: gaps, overlaps and contradictions

Issues

There is a need for a simple, ecologically sustainable approach to conservation.

• Environmental problems associated with land-use intensification have been tackled piecemeal - they have seldom been anticipated or seen as part of a pattern, because legislation is geared to deal with immediate problems rather than ultimate causes.

• Systems of controls and protection that have emerged from this process are therefore complex, and have gaps, overlaps and contradictions. Protective designations also often ignore, and sometimes obstruct, ecological processes.

• There is a growing recognition that ecosystems can respond in complex ways to very simple stimuli. The relationship between environmental damage (complex) and the forces behind land-use intensification (simple) is an example of this phenomenon.

• This relationship between simple actions and complex outcomes may open up the possibility of a simple effective, approach to biological conservation, based on the control of underlying causes.

3.1 Complexity and simplicity in ecological interactions

Biological diversity is the variety of life, and includes variation within and between species, ecosystems and landscapes.

Natural selection, "leads to divergence of character and to much extinction of the less improved and intermediate forms of life. On these principals, I believe, the nature of the affinities of all organic beings may be explained." 'Origin of Species' Darwin 1859.

The relationship between simple forces and complex outcomes is well recognised by ecologists - indeed it is a relationship which lies at the root of biological diversity itself. Natural selection, after all, is basically a simple mechanism - its results, however, are unquestionably complex.

Ecological interactions, in something as mundane as a meadow for example, are highly complex; despite this, species abundance can be altered by something as simple as a change in either fertility, disturbance (mowing/grazing) or both. This can be illustrated by using three types of mesotrophic grassland as examples of ecosystems on a management gradient (Table 11). These are:

- species-rich Crested Dog's-tail dominated grassland (CDT) which constitutes the traditionally farmed meadows still found in some parts of lowland Britain.
- Crested Dog's-tail/perennial ryegrass mix (CDT & PRG) - a widespread grassland derived from a number of grassland precursors through some degree of agricultural intensification.
- species-poor, ryegrass-dominated grassland (PRG) - the intensive, highly specialised counterpart of the Crested Dog's-tail dominated (CDT) grassland.

It is noticeable from Table 11, that within this range of plant communities, intensification is associated with a decrease in species richness (number of species per sample) and a reduction in the number of species that were consistently found in each sample (community constants). There is also an increasing dominance of peren-

Table 11. Relationship between mesotrophic grassland types and management regime.

Grassland types		
Intensive practices (chemical fertilisers/re-sowing) and productivity increasing →		
Hay cut/autumn-winter grazing/dung from animals	Continuous grazing/chemical fertilisers/resown	Sown grassland/chemical fertilisers. Grazed continuously or cut for sileage
Crested dog's-tail (CDT)	Crested dog's-tail and Perennial ryegrass (CDT & PRG)	Perennial ryegrass (PRG)
Community constants	**Community constants**	**Community constants**
Cynosurus cristatus (CDT)	*Cynosurus cristatus (CDT)*	*Lolium perenne (PRG)*
Agrostis capillaris	*Lolium perenne (PRG)*	
Anthoxanthum odor	*Holcus lanatus*	
Centaurea nigra	*Trifolium repens*	
Dactylis glomerata		
Festuca rubra		
Holcus lanatus		
Lolium perenne (PRG)		
Lotus corniculatus		
Plantago lanceolata		
Trifolium pratense		
Trifolium repens		
23 species / sample	**13** species / sample	**5** species / sample

nial ryegrass (PRG). This species needs to be actively managed. It is adapted to fertile growing conditions, and like other competitive grasses, declines in relative abundance when intensive farming practices such as fertilising and re-seeding are removed. This suggests that the simple expedient of reducing farming inputs may allow intensified grassland to return to its derivative type. The rate of reversion would depend upon many factors, not least the dispersal ability of component species and the nearness of source populations. However, as the perennial ryegrass grassland (PRG) reverts via the mixed type of grassland (CDT & PRG) to the one dominated by crested dog's-tail (CDT), there would be a natural increase in structural, spatial and species diversity. This indicates that simple actions, in this case the re-establishment of traditional meadow management, can have complex, but to some extent, predictable outcomes.

At a broader scale, this process is illustrated by the effect of intensive farming on Europe's landscape. Here, increased inputs, principally of agri-chemicals and energy, have resulted in an increase in the fertility of large areas of land. This has reduced the species richness of the affected areas, changed their habitat status and created a more uniform environment.

It seems likely that simple actions can have complex but predictable outcomes at both the level of a biological community and a landscape. This suggests that it may be more effective to control the ultimate causes of agricultural intensification rather than its symptoms. In terms of the grassland communities already discussed, this would mean attempting to reverse a loss in species richness by reducing agricultural inputs, rather than by planting elaborate associations of wildflowers in complex patterns (responding to symptoms). While both approaches may be needed in the early stages of a restoration programme, most ecologists would view a largely symptomatic approach to conservation management as a futile exercise in conservation gardening. Despite this level of awareness, our attempts to conserve the environment tend to be largely symptomatic.

3.2 Attempts to deal with the ecological effects of land-use intensification

Early conservation legislation was drafted to meet worrying circumstances found in north-western Europe, where semi-natural habitats had already been restricted to remnant patches. The effect of this was to put conservation on a siege footing where it became preoccupied with 'protected' sites rather than with the systems and processes that underpinned biodiversity. This approach was born out of ignorance and carries the spirit of defeat - a condition that is reinforced by our continuing inability to protect and maintain designated sites.

Sites of Special Scientific Interest, are the basis for habitat and species conservation in the UK, and are the means of satisfying some international commitments.

The problem with sites is that they are difficult to delineate, neglect important areas and communities that do not fit the sometimes narrow (species orientated) selection criteria, and they tend to be conservative in their land-take. Under current planning regulations, this leaves important dimensions of biodiversity unprotected, and even disadvantaged. For example, in the UK, Sites of Special Scientific Interest (SSSI) are important land-use constraints in the planning process; however, they are merely representative examples of important habitat types, and do not cover the full extent of the remaining quality habitats. This has introduced a sometimes unwarranted differential into the treatment and protection of important natural resources. It is not surprising, therefore, that there has been a move to establish local non-statutory conservation designations to 'fill the gap'. These add to an already complex system of protection and controls and are likely to add to the confusion.

Genetic drift is the random (unselected) change in the frequency of traits in a population of a species. It leads to a progressive loss of variation - but most quickly in small or highly fragmented populations.

The flaw in the site-based approach to conservation is the notion that the long-term survival of something as complex and dynamic as an ecosystem can be secured within a reserve. The reality is that 'protected' habitat fragments are particularly sensitive to their local circumstances, which are often hostile. An assessment of 370 sites 'representative of key habitats' in the European Union (EU), for instance, found that although 70% of the area had some form of protection, almost 60% of the sites were under threat. Although the UK Government thinks that the direct damage to SSSIs has fallen over the last few years, there is little information about how they have been affected by neglect or unsympathetic management, nor about how they are responding to the effect of habitat fragmentation and genetic isolation.

While it is possible that the levels of direct damage to designated sites may decline and be eradicated, as may the effects of unsuitable management practices, fragmentation and isolation will continue to pose serious problems for the development of sustainable management practices. This is because a protection system based on fragments will ultimately encounter the effects of ecological isolation, such as genetic drift, random extinctions and decline unless they are given a favourable landscape setting, or are 'managed' as exercises in conservation gardening.

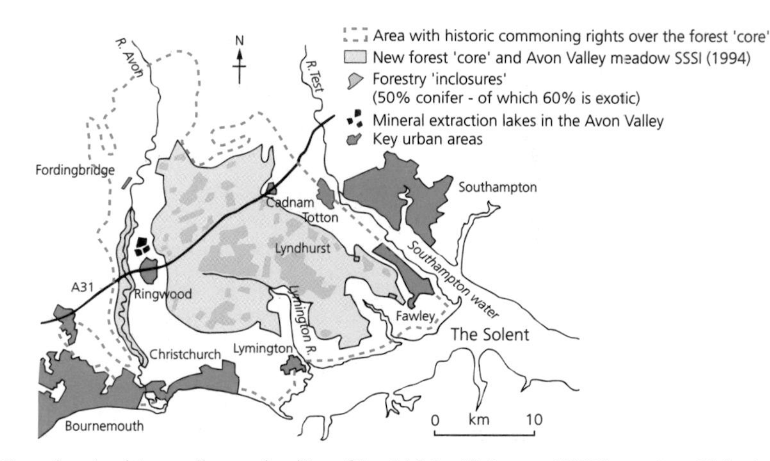

Figure 11. New Forest 'core' and Avon valley meadow Sites of Special Scientific Interest (SSSIs), together with land which has historic commoning rights over the Forest core (sketch plan - not to scale).

Commoning has been widespread throughout Europe for many centuries. New Forest, 'commoners' occupy property with commoning rights over the Forest. Rights include: grazing, pannage (fattening pigs with acorns), turbary (cutting turves for fuel) and estovers (cutting wood for fuel).

3.3 The phenomenon of landscape interactions

Another worrying aspect of site-based systems of protection, is that they are not always able to cater for ecological interactions at a landscape scale. This should not be surprising because sites, by definition, have to be geographically discrete, whereas landscapes, by their nature, tend to be ecologically continuous. This difficulty can be illustrated by some of the relationships within and between the New Forest and the Avon valley in the UK.

3.4 The New Forest, Hampshire (UK)

The Forest is an ancient pasture woodland system that may have its origins in the primaeval forest. By nature, it is a large-scale, shifting, mosaic of woodland, pasture and heathland, where change is mediated by grazing, periodic catastrophes and human influence. The forest core (Figure 11), is protected by an SSSI and surrounded by land that has long-established commoning rights over the core area. This land extends from Hampshire into the neighbouring counties of Dorset and Wiltshire, and includes much of the Avon valley. The maintenance of the forest-core SSSI is heavily dependent on an ancient system of commoning that de-pastures stock into the area from the surrounding land with commoning rights (Figure 12). This land is therefore of some importance to the biological value of the New Forest ecosystem, and presumably establishes an important parameter in the dynamic relationship between the forest core and the surrounding areas.

Pressure from commoning interests and ecological opinion, has encouraged the UK Forest Authority to establish a long-term programme to restore Forest 'inclosures' to their pre-existing habitats (where possible) and open them up to grazing by forest animals

Although the New Forest SSSI was a groundbreaking designation in that it included areas of land that may not on their own have merited SSSI status, it has been unable to protect much of the functionally associated land that has commoning rights over the Forest. Nor has it been able to sustain the commoning system that drives the Forest ecosystem. For many years, this has left both vulnerable to the direct and indirect effects of housing development and mineral extraction. The Forest SSSI has, therefore, been unable to ensure the continuance of the very things that are fundamentally important to the survival of the forest core. Even more remarkable, within the biologically-important core of the Forest, there is a substantial element of forest 'inclosures' (Figure 11) which are managed intensively and have a high proportion of conifers (>50%), many of them

Figure 12. Latchmoor 'drift' - part of the commoning way of life: New Forest ponies rounded up and inspected. Their condition is assessed. Some are sold, those that are in poor condition are 'sent off the forest' to recover on 'backup' pasture. The drift is one of up to thirty that take place in different areas of the forest between August and November each year.

exotic (>60%). These fixed blocks of commercial woodland (Figure 13) not only displace large sections of the ancient pasture woodland ecosystem (Figure 14) with exotic species, but also impede dynamic change. It seems that even with the protection of one of the UK's largest SSSIs, the combined system of environmental legislation and protection has been unable to change the fact that the biologically distinctive core of the New Forest is being compromised from outside and within.

Recently, there has been an attempt to protect a larger area of the land by way of a New Forest Heritage Area. This new layer of control attempts to include enough of the land with commoning rights to ensure the maintenance of the forest core. However, the amount included in the currently proposed Heritage Area is significantly smaller than the area with historically-established commoning rights. This is because the Heritage Area boundary

Figure 13. New Forest near Balmer Lawn, showing forestry planting in Pignal 'inclosure'. The 'ride' separates two ages of planting.

Figure 14. New Forest near Balmer lawn showing the grazed, open structure of the 'Ancient and Ornamental' (pasture) woodland.

was, in part, defined in terms of what was needed to maintain the open forest (the SSSI less the inclosed land). The problem with this arrangement is that it reinforces the status quo in terms of the present levels of 'inclosures'. It also puts the residue of unprotected land with commoning rights at risk - simply because it has (by statute) a lower conservation value than that within the Heritage Area.

3.5 The Avon Valley, Hampshire (UK)

In the case of the Avon valley, the river terraces are part of a traditional system of farming which carries stock that not only maintain the biological value of the valley meadows, but also

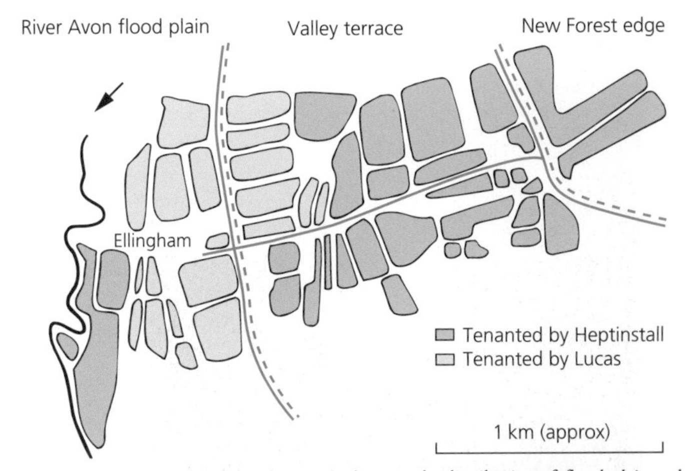

Figure 15. Avon Valley at Ellingham (Hampshire UK) showing the distribution of flood-plain and terrace land among two of Earl Normanton's tenants (George and Ann Lucas, and J.J. Heptinstall) in 1845. Source: 1845 tithe award for Ellingham. Hampshire Record Office accession number: 21m65/f7/77/1-2. Both tenants had land in the valley flood-plain; Heptinstall's allocation was separated from the bulk of his holding by land occupied by Lucas.

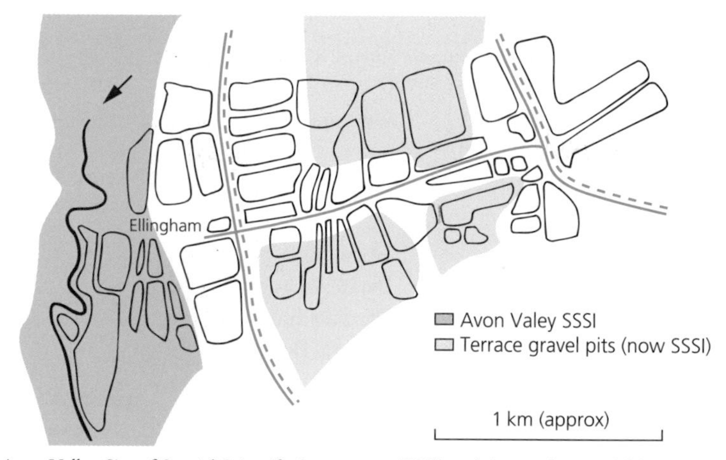

Figure 16. Avon Valley Site of Special Scientific Importance (SSSI) and the nearby gravel lakes (now themselves SSSIs) plotted on the Ellingham 1845 tenancy arrangements shown in Figure (15).

plays a part in the management of the New Forest. Figure 15 shows how within one ownership in 1845, 'parcels' of land were distributed between two tenants. Each 'holding' had an element of both terrace and flood-plain - a not uncommon arrangement in valley farming. Although this functional relationship shaped the ecology of both terrace and floodplain, only the floodplain was given SSSI protection (Figure 16). This discrepancy has meant that while the planning process has been able to 'protect' the river meadow (because it had SSSI status) it has not been able to protect the functionally associated terrace, which has been turned into wet pits by mineral extraction (Figures 16, 17).

The effect of this increasing area of new wet pits has been to erode an element of the valley land-

Figure 17. View from the edge of the New Forest (SU163083) looking west over the Avon valley (Hampshire UK) towards Ellingham. The lake in the middle distance is part of a 1.5 square kilometre system of wet gravel pits between Ringwood and Ibsley. Although they are a new wildlife and recreational resource, they have: a) removed the possibility of establishing the traditional pattern of management that created and maintained the biological value of the adjacent valley meadow-land, and: b) along with other development around the Forest reduced the amount of land with commoning rights.

Table 12. World, European, National and local Conservation Designations arranged in terms of sites, areas and networks. (National and local designations using UK examples).

	Local (UK)	National (UK)	European (EU)	World (UN)
Network			**Natura 2000 series Network**	**Biosphere**
Area	Heritage Areas	**National Parks/Areas of Outstanding Natural Beauty**		**Biosphere Reserves /World Heritage Sites**
Site	Local Nature Reserves/Sites of Importance for Nature Conservation	**Sites of Special Scientific Interest /National Nature Reserves**	**Special Areas of Conservation /Special Protection Areas**	

Designations shown in **bold** have statutory protection. Sites can often be aggregated into areas and areas into networks, and local designations subsumed into those at a national, European and world level.

scape that is important to the maintenance of both the valley meadowland and the adjacent New Forest. The irony of this situation is that some of these extraction areas have gained SSSI status for other reasons such as wildfowl refuges, a fact which mineral operators have used in planning submissions to justify more wet pits.

Both the New Forest and the Avon Valley have been subject to land-use systems that have developed over hundreds and perhaps thousands of years. These systems involve interactions both within and between landscapes, and these interactions have produced the effects which we value in both the Avon Valley meadowland and the Forest. Despite their good intentions, site-based designations such as SSSIs have not been able to foster these interactions to any great extent, and indeed may have frustrated them.

3.6 Protection at a landscape scale

The example of the New Forest and the Avon Valley suggest that site-based approaches to conservation which attempt to protect ecosystems but neglect the underlying causal processes will encounter difficulties in a real situation. When these occur, the stock response has been to make the sites bigger. The current situation is illustrated by Table 12. This shows a complex array of local, national, European, and international conservation designations, together with a number of conservation-related designations such as National Parks and Areas of Outstanding Natural Beauty. Generally, sites can be subsumed into areas and areas connected as networks, at a national, European and world scale; in each case, with an increase in size and a greater commitment to large-scale management.

Some EU Member States, have already used the available legislation to create biologically appropriate areas. They have done this by connecting Special Protection Areas (SPA) (Article 4 of the EU birds directive) and Special Areas for Conservation (SAC) (Article 1 of the EU

Agri-environment schemes are locally applied elements of the Common Agricultural Policy that provide financial incentives for farmers to manage land in an environmentally-friendly way

Agenda 21, is a United Nations declaration that recognises the interdependence of ecological processes and the need for a sustainable relationship between man and the environment.

habitats directive) within an EU-wide "Natura 2000" network (article 3 of the habitats directive). These initiatives have introduced the benefits of scale. They are also potentially supported by a number of environmental measures under the Common Agricultural Policy (EC regulation 2078/92), including a range of agri-environmental initiatives. However, it is difficult to argue that they are anything more than a site-based approach writ large, because they do not address causal issues in the form of land-use intensification. In some instances, they are not even this useful - the UK Government, for instance, fulfils its EU commitments to landscape-scale designations by requiring the prior designation of SPAs and SACs as SSSIs. Given the fragmented nature of SSSIs and the inexorable land management pressures working against them, this somewhat defeats the object of the exercise.

Conflict between modern land-use management and site-based approaches to conservation, has lead to an increasingly complex system of protective designations, which, on the 'Russian doll' model, has attempted to handle problems by deploying progressively more extensive (and arguable more confusing) legislative shells. What this approach largely ignores, however, is that a healthy, robust, and biologically-rich environment is the outcome of a sustainable relationship between man and the land (essentially the message of Agenda 21). In taking this line, it becomes clear that protective designations can generally only have value for limited time periodor as part of a wider strategy, and that the only realistic long-term

Info Box 4: The European Union (EU) Biodiversity Strategy

The 1992 Convention Biological Diversity (CBD) placed an obligation on each signatory to prepare a biodiversity strategy, and within this framework, elaborate the required action plans (Article 6).

The EU Biodiversity Strategy was published in 1998. Its main thrust is to provide a coherent framework for the conservation of biodiversity throughout the Commission's activities. This will be attempted by integrating conservation targets into a range of sectoral and cross sectoral policy areas. To achieve this, the strategy sets out a number of 'themes'. These themes have conservation aims which are derived from the CBD, and these aims must be addressed by specific 'objectives' within each policy areas or 'sector'.

For example, one of the aims of the 'Sustainable use of biodiversity' (Theme 2), is to support the social and economic viability of systems that underpin biodiversity. In the case of agriculture (Sector 2), the corresponding 'objective' is to continue shifting agricultural support away from intensive farming practices towards those that enhance biodiversity (extensive systems). In the case of 'Regional policies and spatial planning' (sector 4), the relevant objectives will relate to the design of zonal programmes (reflecting the natural distribution of landscape types) and the targeting of financial instruments such as the Structural Funds, the Cohesion Funds and the efforts of the European Investment Bank.

The level of inter-linkage between thematic aims and policy area objectives across the sectors has the potential of taking conservation policy to the heart of mainstream land-use and economic planning. It needs to be remembered, however, that the strategy is simply a framework that will need to be elaborated by preparing, implementing and refining the relevant action plans. These are currently being developed (1999) and, hopefully, will be put into practice within two years.

Themes

1. Conservation of biodiversity
2. Sustainable use of biodiversity
3. Sharing benefits of the use of genetic resources
4. Research, monitoring and exchange of information
5. Education training and awareness

Sectors (policy areas)

1. Conservation of natural resources
2. Agriculture and forest policies
3. Fisheries
4. Regional policies and spatial planning
5. Transport and energy
6. Tourism
7. International co-operation

3

approach to biological conservation is by way of a sustainable system of land-management.

3.7 An ecological approach to conservation and land-use management

In order to maintain and restore some of Europe's regional diversity of wildlife and habitat, it is likely that conservation measures will need to be integrated into the way we manage land. This implies a knowledge about the natural distribution of Europe's habitats and landscape types, an understanding of their natural carrying capacity, and a willingness to manage them appropriately within the context of a sustainable system of land-use. The possibility of this level of integration is being examined at a number of levels across Europe, especially in those areas concerned with spatial planning, strategic impact assessment and policy formulation.

Carrying capacity is the ability of an ecosystem to be productive while remaining adaptable and capable of renewal.

3.8 Spatial Planning

The process of defining and mapping the natural distribution of landscape types and potential habitats has already been started at a number of administrative levels across Europe. There have also been some attempts to stratify agricultural landscapes according to sound agricultural and ecological criteria. Most of the attempts have been very rudimentary and there are some differences in approach - even within member states; however, these various initiatives indicate a common resolve to define and map Europe's landscapes as a basis for assessing the impact of strategic policies on the environment.

The first draft of a harmonised framework for the classification of European landscapes is in its consultation stage (1998) and could be implemented by 2000.

3.9 Strategic Environmental Assessment

Environmental Impact Assessment (EIA) has been around for some time and, though its scope has been limited, and the ecological content of many impact assessments very basic, it has focused minds on the need to take a broad view of development proposals. Recent legislative changes in Europe reflect the concern that there is an increasing need to assess environmental impacts at a strategic level. It is hoped that these Strategic Environmental Assessments (SEA) will be a the means of evaluating the long-term environmental impacts of strategic proposals at different levels of the policy making process. Although take-up of the SEA approach is patchy between, and even within, member states, it should ultimately help to integrate environmental considerations into the policy making process and give early warning of proposals that may damage the environment.

3.10 Policy Integration

Policy integration or 'joined-up government' still has a long way to go in Europe. Nevertheless, steady progress is being made under the provisions of the internationally ratified vention on Biological Diversity (CBD). An important example of this is the recently published (1998) European Union (EU) biodiversity strategy, which gives the process of integrating conservation and land-use practice a high priority (InfoBox 4). Interestingly, the strategy establishes the link between high biodiversity and traditional systems of land-management, and also highlights the need to support these systems, together with their indigenous communities, related technologies and domesticated species. These, and other concerns throughout the EU biodiversity strategy, seem to put traditional land-use systems close to the centre of the European Commission's thrust to meet its obligations under the 1992 Convention on Biological Diversity. It will be important to see how these will be put into practice.

These developments represent a significant coming together of the legislative instruments needed to bring about an ecological approach policy integration. This can be shown in relation to European agricultural policy. At the moment, serious environmental problems still occur because plans, policies and programmes are generally unable to reflect local environmental differences (Policy Integration). This is because differences in the character and carrying capacity of the land have not been comprehensively characterised or quantified (Spatial Planning). However, even if they had been characterised, there is no established mechanism for assessing the impact of policies, plans and programmes on different landscape types (Strategic Impact Assessment).

3.11 Concluding thoughts

Lessons from our recent history indicate that intensive commercial production is not a universally sustainable option, and future land-use policies will need to reflect this fact. Future production systems will need to work within the natural limitations of the land, much in the way locally-adapted systems of extensive farming have done for centuries. Despite the impact of intensive land-management over the last fifty years, remnants of these ancient and well proven systems of production still remain, along with their indigenous technologies and community structures.

In the case of the New Forest, the system has survived the vagaries of political and agricultural change more or less intact for over a thousand years. The last few decades, however, have brought increasing social, economic and land-use pressures, and these have left the Forest ecosystem in a vulnerable state. While designations may have 'protected' some parts of the Forest system by isolating them from change, they have also had the effect of consolidating inappropriate land uses, such as commercial plantations, and they have rendered some unprotected areas vulnerable to loss and degradation. It has been known for a long time that the survival of the forest ecosystem does not depend upon protective designations, but upon our ability to continue the underlying process that have formed it over the centuries (Info Box 5). Ultimately, the same is true of other ecosystems throughout Europe.

In thinking about Europe's surviving systems of traditional land management and their association with rich natural habitats, it is difficult to escape the conviction that much of what we prize in the landscape is the outcome of systems of land-use that worked within natural limits. There were no 'experts', or 'earnest societies' involved, or protected sites - just the enduring application of locally developed crops, animals and technologies to human need. The next chapter explores what we know about such systems.

Info Box 5. Farmers or park keepers: ecosystem management or landscape gardening

"If you would preserve rural England preserve the rural Englishman and his means of livelihood. Otherwise all your earnest societies for this and that are but pious futilities, setting out to 'freeze' a landscape to a status quo the factual validity of which has departed, gone with the population and the folk-life which were its roots before they withered away. The artificial preservation of such a simulacrum *in vacuo* is a vain imagining. Such landscape gardening of all England would demand hordes of 'experts'. park keepers, and paid labour such as would beggar even the wealth of England in 1913."

The Commoners New Forest (1944)

3

Summary:

The complexity of habitats and landscapes that were once characteristic of Europe is the result of farming within the natural constraints of the land (a simple determinant). Similarly, the complex and unwanted environmental effects of modern land management are the result of policies that encouraged farmers to exceed these constraints (a simple determinant).

Conservation has not tackled these simple causal mechanisms but has been preoccupied with their rather complex effects. Over the years, this has generated an equally complex system of legislative controls and site-based protection which has been largely unable to prevent or reverse losses. In some cases, they may even have made the situation worse.

Given the relationship between simple actions and complex outcomes, the most effective way of dealing with many of our current environmental concerns is to farm within the constraints of the local environment, using as a basic model, traditional, locally adapted farming systems.

4 Traditional land-use systems: sustainability, efficiency and biodiversity

Issues

Currently used methods of intensive land management are unsustainable (Section 2), and our attempts to protect the environment from their damaging effects have at best been limited, and in some cases counter-productive (Section 3).

Robust, biologically-rich landscapes seem to be the natural result of using locally adapted land-management systems. Although these are not as productive as intensive systems, they have a rich variety of produce, and are sustainable, efficient and capable of providing the basis for biological conservation at a number of ecological scales. Unfortunately, they are now largely confined to the least versatile soils, and are declining.

4.1 Sustainable land-uses

Although there are many definitions of sustainable use, well-tested models of sustainable practice have been around for a long time, and still survive in the form of locally-adapted, traditional low-input land-use systems (Info Box 6). These offer a sound basis for sustainable land-management while creating and maintain high levels of biological, cultural and economic diversity. With the right framework of support, they have the ability to flourish and provide sound models for sustainable land management systems throughout farming and forestry.

The inherent sustainability of low-intensity pastoral systems is linked to the fact that they are a human adaptation of conditions that already existed in the post-glacial forest landscapes that were grazed by large herbivores. Indeed, the many transhumance systems mimic the seasonal movement of grazing herbivores.

Although a conventional palaeo-ecological view is that much of Europe was subject to a neat progression from tundra to forest, the reality is likely to have been more chaotic and diverse, with the role of large herbivores and open conditions playing a more important part than previously thought. It is likely that the early models for extensive land-use would have been shifting, multi-scale tapestries of woodland, heathland, grassland and wetland in which the balance of habitat types was strongly influenced by grazing vertebrates. This would help to explain why, despite the short time frame, the behaviour and form of much of Europe's wildlife is adapted to open conditions. That is not to say that extensive systems of land-management have not interacted with the underlying conditions to enrich biodiversity: they clearly have; locally adapted races of crops and domestic animals are good examples of this continuing process. This is partly why traditional systems of land management are so important ecologically.

Although, within living memory, much of Europe's landscape was farmed extensively with only minimal external inputs, less than

4

Info Box 6 Has European agriculture ever been sustainable?

The long association of European wildlife and pastoral or mixed agriculture is often overlooked. Ten thousand years ago, in much of Europe, open forests, grasslands, heaths and marshes began replacing the ice-age landscape. Three thousand years later (around 5000 BC) the forests were already being cleared further by Neolithic people and the naturally-occurring large herbivores and carnivores, that were later replaced by domestic livestock.

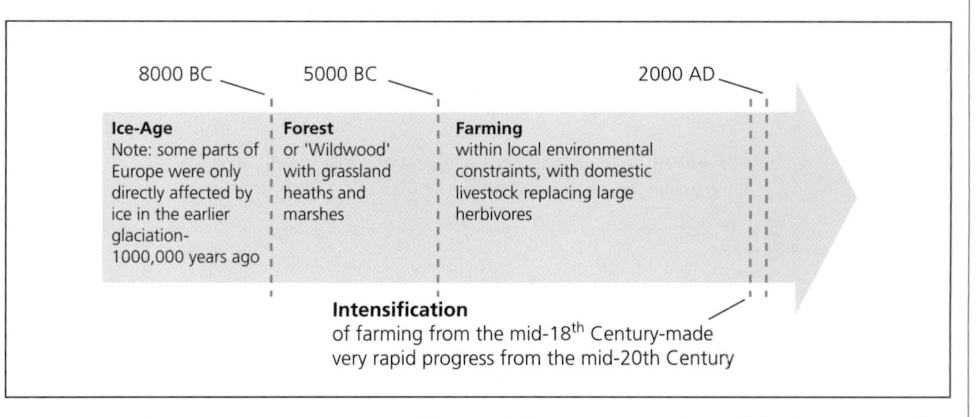

Europe now has a managed landscape. This evolved over a period of almost twice as long as that occupied by the post-glacial, forest, grassland and heath vegetation. The present day grassland, heath, moorland and bog are in part the result of farming systems, which through to the mid 18th century, developed highly integrated regional livestock farming practices, with indigenous technologies and locally distinct breeds of sheep, pigs, cattle and horses.

Were these good models for sustainable land-use systems? There are many definitions of sustainability. However, a practical demonstration of sustainability is provided by Europe's many systems of extensive pastoral land-uses which involve strong linkages between biological, ecological and social factors. Many of these have kept going for as long as 7000 years, supported over 300 generations of people without significant external inputs, and, until a few decades ago, carried rich populations of wildlife.

40% (56,000,000 ha) of the 'utilisable' agricultural land is now managed this way (Figure 18). This is largely restricted to the least productive landscapes (Figure 19) and represents the declining residue in an ongoing process of intensification and neglect.

4.2 Character of extensive land-uses

Low-intensity farming systems across Europe have many things in common, but they also show a wide variety of local variation. This should not be surprising because they conform to and enrich a landscape that is characterised

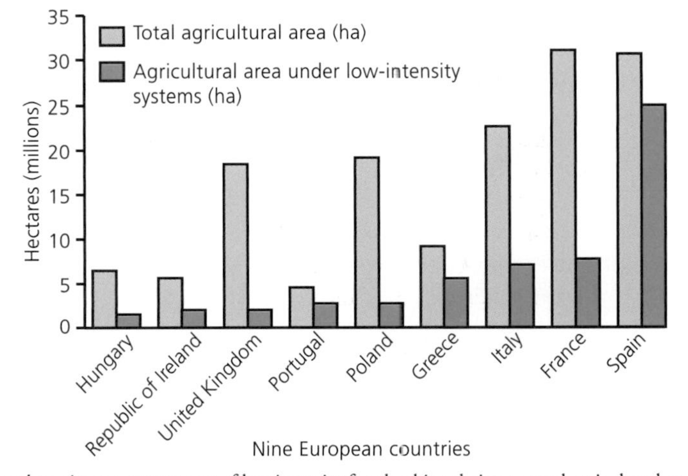

Figure 18. Estimated maximum current area of low intensity farmland in relation to total agricultural area of nine European countries.

by a complex range of ecological conditions. The typical characteristics of 'low intensity' systems are summarised in Table 13. They tend to have few additions in terms of basic infrastructure such as drainage and irrigation, and have high amounts of semi-natural vegetation. They also have few inputs of animal feed and agrichemicals.

Fertility and food supply, therefore, have to be maintained by adjusting the relationship between woodland, pasture, crops and animals, often within a rotation or by seasonal exploitation. For example, the transhumant sheep of Iberia deposit large quantities of dung across areas which would otherwise maintain few livestock. In this way, the sheep constitute a transient means of converting vegetation into natural fertiliser. This kind of sustainable land management, limits the use of external inputs, maintains soil structure, reduces nutrient leakage into the wider environment, and minimises

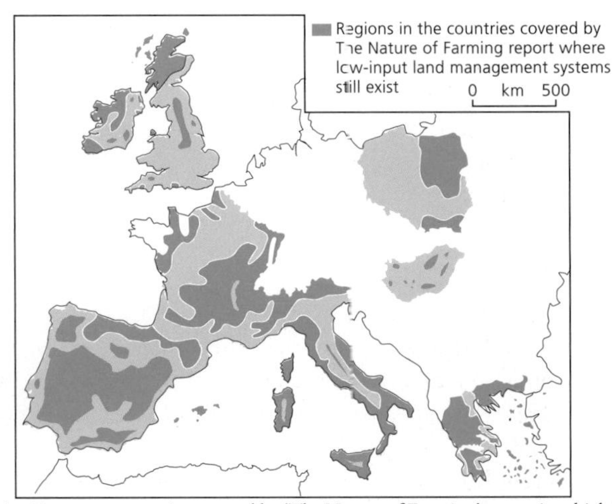

Figure 19. Regions of the nine European countries covered by 'The Nature of Farming' report in which low-input systems are potentially still common.

Table 13. Typical characteristics of low intensity systems. *Nature of Farming 1994*

Livestock systems	**Crop systems**
• low nutrient input, mainly organic • low stocking density • low agri-chemical input • little investment in land drainage • relatively high level of semi-natural vegetation • relatively high species-richness of sward • low degree of mechanisation • often hardier regional breeds of stock • survival of long-established management practices, e.g. transhumance, hay making • reliance on natural suckling • limited use of concentrate feeds	• low nutrient input, mainly organic • low yield per hectare • low agri-chemical input (usually no growth regulators) • absence of irrigation • little investment in land drainage • crops and varieties suited to specific regional conditions • use of fallow in the crop rotation • diverse rotations • more traditional crop varieties • low degree of mechanisation • tree crops, tall rather than dwarf - not irrigated • more 'traditional' harvesting methods

problems with pests and diseases. It also results in the cultural, economic and ecological linkages that make them potentially more sustainable (and diverse) than high-input commercial systems of land-use.

The mutual dependencies within extensive systems embrace relationships that extend from the simple plant-animal interactions that give habitats their distinctive structure, through to the indigenous land-use technologies and cultural systems that create and maintain whole landscapes. Some of these dependencies can be illustrated by the biologically rich dehesa (Spain) (Figure 20) and montado (Portugal) (Figure 21) pasture-woodland systems that cover more than three million hectares of Iberia. These are open mosaics of pasture,

Figure 20. Spanish dehesa, Extramadura. Woodland of Holm Oak and Cork Oak. Van Dijk, 1994.

Figure 21. Montado in the Alentejo region of Portugal, showing cereal cultivation under open woodland. Natacha Yellachich, 1992.

woodland, scrub and aromatic shrubs, with patches of arable land on the more fertile soils.

The pattern and structure of the dehesa mosaic, and thus the character of its dependent habitats are shaped and maintained largely by highly integrated systems of extensive farming. For example, the dominant cork oak (*Quercus suber*) and the holm oak (*Quercus ilex*) are important to the local economy for cork and acorns. The latter are used to fatten the native black Iberian pig (Figure 22) which produces the distinctive, and highly marketable regional hams (e.g. "pata negra"). In order to maintain acorn production, the holm oak are traditionally pruned on a 10-20 year cycle; these prunings yield a number of by-products which include firewood, saleable charcoal, and leaves which are used as animal fodder.

Although the perception of 'modern' farming is that low-input practices are low-yielding

Figure 22. Black Iberian pig (some are cross breeds) grazing the montado in the Alentejo region of Portugal. Gwyn Jones.

Table 14. Relative efficiency of intensive and low intensity farming systems taking into account their wider benefits and penalties in terms of inputs and outputs; H = High, L = Low.

	Farming system	
	Intensive	**Low-intensity**
Input costs:		
Fuel	H	L
Agri-chemicals	H	L
Water	H	L
Labour	L	H
Feed	H	L
Public funds	H	L
Transport costs of inputs and products	H	L
Ecological and landscape simplification	H	L
Risk to gene-pool of domestic breeds	H	L
Erosion of species barriers	H	L
Epidemiological risk	H	L
Pollution of air and water	H	L
Soil erosion	H	L
Outputs		
Food and other agricultural products	H	L
Maintenance of landscape and ecological complexity	L	H
Locally distinctive patterns of production and quality products	L	H
Locally adapted crops and livestock	L	H
Maintenance of rural communities	L	H
Landscape character and quality	L	H
Potential for sensitive tourism and other sustainable rural development	L	H

anachronisms, the dehesa example shows that they can be highly efficient means of exploiting the naturally available resources. In the case of intensive production, however, it is difficult to confirm that the value of outputs in terms of production are greater than the inputs in terms of fertilisers, feed and public money. When the wider benefits and penalties of the two systems of farming are assessed (Table 14), it becomes clear that not only is low-intensity farming an efficient and sustainable means of production but that it has more benefits and fewer penalties than the intensive alternative. By contrast, the 'efficiency' of intensive production is not only questionable; it exacts heavy environmental penalties - one of these, soil erosion, even undermines long-term basis of production (Section 2). This suggests that the worst examples of intensification are not only inefficient but also self-destructive.

Unfortunately, land-use systems are not organised around the non-food outputs listed in Table 14, but focus on short-term economic considerations. A problem with this is that the economics of intensive land-management practices are narrowly production-orientated and thus neither cost their long-term damaging impacts nor recognise the wider benefits of traditional land-management systems.

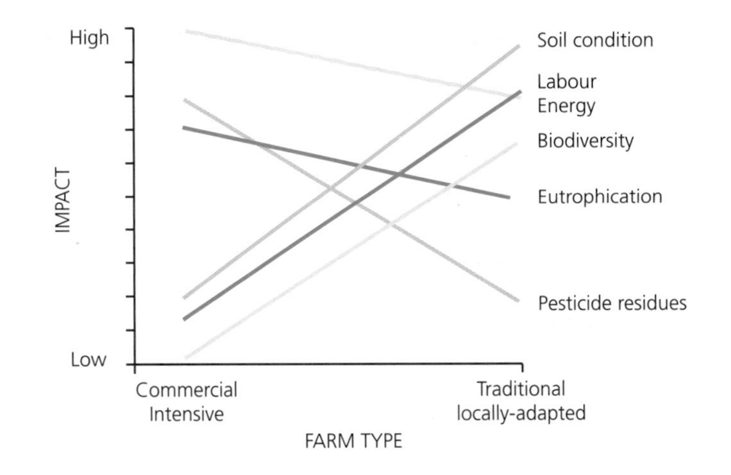

Figure 23. The environmental impact (high/low) of different farming systems (after Gun Rudquist: The Common Agricultural Policy and Environmental Practices, 1996)

When considering the relative efficiency of land-use systems, what should really concern us is not their crude ability to increase production; instead, we should be interested in their ability to yield useful products with minimal external inputs, and without harm to the ecosystem. This approach relates well to the concept of sustainable use, and to the elegance of ecological management; however, it challenges current cultural and economic thinking. The purpose of formal agricultural training, for instance, (with considerable input from the vested interest of the agri-chemical industries) would suggest that success in farming involves drastically changing the natural environment, or in the case of the plants and animals, their genetic make-up. Ecological management, however, would argue that the mark of success, and indeed the real technical and scientific challenge is to have the skill and knowledge to get the natural environment to work for you. This approach is at the heart of sustainable land- use.

Figure 23 shows how a range of environment variables such as soil condition and biodiversity are affected by land management practices. It is noticeable that the complex processes associated with lowering pollution and improving soil condition and biodiversity are accomplished by the simple change from intensive (high input) to low intensity farming (some types of organic farming being one example of the latter).

4.3 Farming efficiency and ecological complexity

As the Dehesa example showed, low-input methods of land-use can be remarkably efficient - in the true sense of output per unit input. Throughout Europe, traditional, locally-adapted systems of agriculture have shown themselves to be highly adept at exploiting sometimes transient commercial opportunities, both within and between ecosystems. Interestingly, the relationship between human exploitation and high biodiversity in extensively farmed areas, suggests that within the 'low-input' paradigm (Table13), the process of maximising commercial exploitation increases biodiversity. Domestic breeds of animal and crops are an obvious result of this process. However, this phenomenon is also evident at a variety of levels of ecological organisation.

4.4 Habitat scale effects of optimising production within extensive systems

At a habitat scale, for instance, most extensive systems use a range of animals to exploit differences in the state and type of vegetation. Sheep, cows, pigs, and horses have different grazing patterns and preferences. Used separately, as mixtures or in sequence, they help to maximise productivity, and in the process, create and maintain extensive vegetation mosaics at a range of ecological scales. These are important for wildlife because they provide a range of conditions in which natural ecological processes can take place.

For instance, on Islay in the Inner Hebrides (Scotland), the efficient use of complementary stocking patterns within an extensive low-input pastoral farming system is important for a number of wildlife species including Chough, Golden Eagle and wintering Geese. This richness of conservation interest has been shown to be associated with a farming system that involves a continuity of grazing management, the growing of winter fodder, and the use of native highland cattle (Figure 24).

Like other native breeds, the highland cattle consume large quantities of rough, nutritionally-poor vegetation and need little supplementary feed. The grazing pattern of this herd is more like that of wild herbivores than modern domestic breeds, and so the vegetation mosaics they create are more varied and have a richer wildlife. The ecological importance of this traditionally derived variation of suckler cow management, and its effects on the pattern, structure and species composition of farmland can be shown by looking at two rare but locally abundant species, the Marsh Fritillary butterfly *(Eurodryas aurinia)* and the Red-billed Chough *(Pyrrhocorax pyrrhocorax).*

4.5 Marsh Fritillary butterfly *(Eurodryas aurinia)*

The Marsh Fritillary (Figure 25) is the butterfly species most threatened by extinction in Europe and has its strongest populations in the UK. It breeds in two distinct types of habitat: damp, neutral or acid grassland, and dry calcareous grassland, but in both habitats the

Figure 24. Croftland and Farmland in Islay, Inner Hebrides. A mix of pasture, meadow, cultivated land and woodland. Roger Wardel, 1995.

Figure 25. Marsh Fritillary butterfly, adult. A Campbell.

foodplant, Devil's bit scabious (*Succissa pratense*), is the same. The butterfly is on the wing for only a short period during May or June when the temperature is high enough and wind speed low. Adults lay their eggs on the underside of the leaves of the larval foodplant, where three weeks later they hatch to form colonies of caterpillars. Soon after emergence the caterpillars spin a dense web over the foodplant, beneath which they live and feed, moving to another plant when the first is consumed (Figure 26). During August, a more substantial

Figure 26. Marsh Fritillary laval webs on sheep and cattle grazed coastal pastures with *Succissa pratense*, Islay. Eric Bignal, 1996.

web is constructed, inside which they hibernate, emerging the next Spring to bask on sunny days and recommence feeding. They eventually disperse from the colony having spent about ten months in the larval stage. The caterpillar pupates close to its food plant and hatches about a fortnight later.

In Britain the Marsh Fritillary butterfly is a species of the heath and acid grass pastures of western Wales and western Scotland. In Scotland the height of the vegetation where caterpillars are found in mid-summer, ranges from 11-35 cm, but 80% of sites are within a much narrower range (11-14 cm). These heights are significantly different from a random sample of vegetation heights in these areas, showing that the butterflies select this height sward preferentially.

The vegetation within which eggs and caterpillars are found occurs on a mosaic of wet heathland and grassland which is maintained by the extensive grazing of free-range cattle. All UK colonies show similar attributes with the optimum conditions being a summer sward height of 4 -14cm with abundant food plants, in an area where the vegetation structure gives the butterfly and the foodplant shelter without dense shading. These conditions are provided on pastures where the vegetation of grass and heath is maintained by the routine seasonal grazing pressure of cattle and sheep managed in an extensive system. Sheep grazing alone, supplemented by periodic burning (a common management regime today requiring less labour) does not provide suitable conditions, instead leading to heather monocultures and short sheep-grazed grassland.

In areas with optimum grazing pressure for this species, the low density of grazing animals, primarily cattle, ensures that the risk of livestock trampling the colonies of caterpillars is low, yet at the same time the grazing pressure is high enough to remove the coarse vegetation growth, favour the foodplant and produce the optimum height of vegetation.

The species typically undergoes large fluctuations in population size and colonies regularly reach low levels. This may be a result of parasites, changes in habitat suitability, weather conditions or periods of inappropriate management. However, since the species is relatively mobile (with a colonisation range of 15-20 km), the nature of the intervening land between the core areas may be crucial in determining the success or otherwise of recovery after population declines. In other words, the long-term future of populations may depend as much on "sub-optimal" habitat (which is only occupied in "good" years) as on the core areas.

In this respect Marsh Fritillary butterflies colonies are likely to be interconnected and may function as meta-populations - a collection of local populations, connected by occasional dispersal, in which there are local extinctions and recolonisations. The main implication of this is that the long term survival of the species will be determined by the survival of large areas of pasture land grazed by cattle at low intensity, rather than the management of places where they are currently colonising. Indeed in many regions (e.g. in Berkshire and Oxfordshire, UK) the butterfly is probably already doomed because colonies have become too small and too isolated to persist for much longer.

4.6 Red-billed Chough *(Pyrrhocorax pyrrhocorax)*

The Chough (Figure 27) is a rare bird with a fragmented distribution in Europe. Over half of the estimated minimum European population of 16,000 pairs is concentrated in Spain, Greece and Italy; most populations are small

Figure 27. Chough adult with nestlings, Islay. Nestlings have individually coloured rings to aid identification after fledgling. Martin Withers, 1984.

and many are declining.

Choughs feed principally on soil-living, surface-active and dung-associated invertebrates which they obtain from extensive pastures grazed by sheep, cattle, horses and goats. Optimum feeding conditions are bare ground and short or open vegetation in a wide range of arid to temperate climates. However studies in the Scottish Hebrides have shown that there are:

- different foraging strategies between breeders and sub-adults and non-breeders;
- seasonal shifts in diet through the year, and;
- critical periods during which mortality exceeds recruitment.

The Scottish study also found that the seasonal abundance and availability of the main items in the chough's diet are influenced predominantly by agricultural management:

- cereals are obtained from stubble fields, cattle feeding stations or cattle dung (part digested) from October to April;
- fly larvae are taken from pastures between January and April, and in large numbers in July;
- insects are obtained from cattle and sheep dung during spring and in late summer and autumn;
- during high summer, a wide variety of surface invertebrates are consumed from the pastures.

During the periods of incubation and nestling rearing, the breeding birds select grass pastures preferentially to other habitats for foraging and feed predominantly on soil invertebrates - for example, Crane fly larvae. The abundance and availability of Crane fly larvae in the soil were found to be related to the agricultural management of the fields. The fields preferred by the birds in 1992 were those in which:

- management during June to September 1991 produced medium to high grass swards, either through a low stocking density of cattle or the growth of silage or hay crops for late mowing, and;
- intensive grazing by sheep and cattle between January and May 1992 reduced the vegetation markedly.

The former management produced optimum conditions for adult Crane flies to oviposit (in

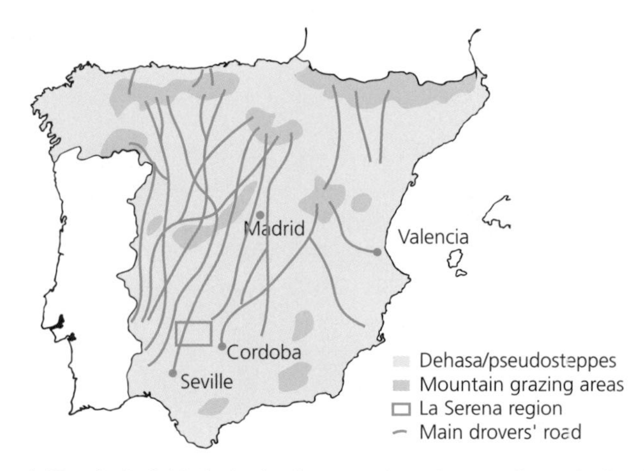

Figure 28. Main drovers' roads (Cañadas Reales) in Spain, showing mountain grazing areas. Source: La Cañada, March 1994.

the medium to high grass swards) and the latter management created optimum feeding conditions (short vegetation, bare ground) for Choughs foraging for Crane fly larvae.

Weather conditions in autumn and spring affect Crane fly populations but are largely beyond control; however, in many years, the manipulation of pasture management can potentially create optimum conditions at a critical time and have a beneficial effect on nestling survival and subsequent population recruitment. Although the ecological conditions needed for this rare European bird are complex, they can be effectively provided by routine farm management, timed to produce the desired conditions.

Common requirements of the Chough and the Marsh Fritillary. The examples show that the practical application of a traditional locally-adapted system of land management can create a habitat-scale complexity that is essential for two species with very demanding ecological needs.

The survival of these species, therefore, relies on a type of ecological complexity produced by the simple routines of extensive, low-input farming.

4.7 Landscape scale effects of optimising production within extensive systems

The commercial exploitation of the land by low-input farming systems, however, not only creates diversity within and between habitats at a local scale; it also maintains a rich pattern of biodiversity across whole landscapes. The surviving practice of transhumance in southern Europe, for instance, involves the movement of hardy locally-adapted stock between lowland (winter) and upland (summer) regions. This exploitation of seasonally available pastures not only helps to maintain commercially viable stock numbers, it maintains a range of distinct landscapes - often separated by great distances.

The historic importance of transhumance is shown by the extent of drovers' roads (Cañadas Reales) in Spain (Figure 28). These have linked the lowlands grasslands and pasture woodlands with the mountain pastures for more than eight hundred years, and in the eighteenth century carried an annual movement of three and a half million animals. Along with similar systems in France, Greece and Italy, however, Spanish transhumance is declining and threatens to follow a loss of similar systems in north-western Europe.

Over much of southern Europe, the decline of transhumance is associated with a pattern of overuse and neglect. Most of the overuse is taking place in the lowland areas, with only localised overuse in the more accessible upland areas. This continuing process of land-use polarisation is simplifying upland and lowland ecosystems, reducing the abundance of wildlife species associated with high levels of biodiversity and making landscapes more uniform.

The overuse of the steppe land in the La Serena area of the Extremadura, Spain (Figure 28), for example, has caused a loss of soil organic matter, localised nitrogen pollution, and a proliferation of fencing. Overstocking has resulted in damaged pastures, the removal of cover and food for birds, and has resulted in the trampling and disturbance of nests and young birds in the spring. The impact of these changes on wildlife has been dramatic. Between 1986 and 1993, there was a significant decline in some specialist steppe birds such as Stone Curlew(75%), Little Bustard (75%) and Golden Plover (50%). These trends seem to be common to the whole of the Iberian pseudo-steppes, where losses have been attributed to an increase in such things as: irrigation, afforestation, scrub regeneration, cultivation, pesticide use, and changes in stocking densities.

Less is known about the effects of the decline of transhumance on Europe's mountain pastures, but transhumance has been important in creating and maintaining their distinctive ecological make-up. In the Pindos mountains of Greece, for example, transumance not only supplies the pattern of short-term, intensive grazing needed to maintain the botanical interest of the pastures above the tree line, but it also provides carrion for a number of important bird species, including the Golden Eagle (*Aquila chrysaetos*). In some areas, such as Valle d'Aosta, Italy, short distance transhumance also contributes to biodiversity below the tree line because of the need for fodder production from the mid-altitude hay meadows.

4.8 Concluding thoughts

The habitat and landscape effects of extensive land-use management are not independent because the cross-landscape impact of transhumance involves very precise habitat-scale effects. For instance, the short-term, intensive grazing in the mountains and the seasonal remission of grazing in the lowlands have profound implications for the species composition and structure of each system, and thus the survival of their dependant species. This highlights the point that complexity at a number of levels of ecological organisation is produced and maintained by (relatively) simple and routine farming operations that have evolved within locally-adapted systems of low-input of farming.

Although low-input systems may have produced much of what we value in our environment, they are maintained by a balance of social, environmental and farming influences and are susceptible to changes in these areas. The effect of agricultural policy over the last fifty years is testimony to this, but so is the effect of the more general, background changes in the way we live. For example the neglect and intensification that is occurring in the Spanish Dehesa is not simply the result of agricultural policy, but has also been influenced by such things as the fall in charcoal prices due to a more ready supply of electricity and the move from rural to urban employment. It follows from this that a switch in policy support from intensive to low-input systems may not work unless efforts are made to adapt and re-equip their social and economic infrastructure.

Summary:
Low-input systems of land-management maximise productivity within the limitations of local conditions; they are both efficient and sustainable. This business-like relationship between man and the land (a simple determinant) has produced an environment which is not only rich in landscapes, habitats, wildlife, and locally adapted breeds of crops and stock, but is also rich also in its indigenous cultures, land-based technologies, and high quality produce.

Such are the complexities of these landscapes and their dependent social and biological systems, that it is difficult to see how they could be designed or recreated. This is why those that remain should be protected from further 'neglect' and decline, and why attempts to restore those that have been lost should proceed slowly, through the 'natural' interaction of local circumstances and tried 'low-input' management practices.

A key feature of this process of reform would be a shift in support from intensive to low-input farming practices, thereby allowing agricultural productivity to be determined by local circumstances rather than by inputs of fertiliser, energy and market support. This would have the effect of freeing-up farmers to farm their land in a sustainable yet efficient way, and will, in time, restore the full richness of Europe's countryside.

5 Land-use reform: a new harmony between human activity, economic forces and the land

Issues

A European route to sustainable land-use management.

- In western Europe the largely sustainable relationship between human activity, economic forces and the land, has been radically altered over the last fifty years, and so has the fabric of the landscape and its rural communities (Section 2).
- This change has been fuelled by policies which rewarded perceived needs both to increase supplies of domestically produced foods and to maintain farm incomes. These policies have incorporated a variety of direct and indirect support mechanisms which have favoured or 'weighted' production by fixing prices above free-market values.
- The challenge now is to use European land-use policies to re-establish a harmony between human activity, economic forces and the land. This will help to secure a sustainable system of production, and provide and a good quality of life for farmers and good quality products for consumers.

Although European art and literature have given us a romantic view of the countryside, this should never mask the reality that the managed landscape is a product of hard, unrelenting economic forces, and a place where people make a living. Until about 50 years ago, production had traditionally been regulated by the interplay of market forces and natural land-use constraints, or by cultural controls that took account of these interactions. Generally, when the price was right, farmers intensified production, but rarely by over-riding the sometimes subtle restraints of soil, land-form and climate. This very prosaic economically-based relationship between human activity and the landscape has moulded Europe's rich tapestry of landscapes and cultures.

5.1 Agricultural support mechanisms

The way in which post-war subsidies have affected production and thus the environment can be seen by considering livestock production. The European Union (EU) Common Agricultural Policies (CAP) subsidy for livestock has two components. These may operate together or alone, depending on the system of production being supported. Generally, there is a direct payment for each animal (headage payment) and an indirect payment for the output of each animal in terms of milk or meat (price support). These mechanisms result in different levels of support for each type of production, which increases from sheep, to beef, through to dairy (Info Box 7).

Across Europe, these incentives have stimulated production and created surpluses - both contributing to a soaring CAP budget. They have also brought about a simplification of the landscape through a combination of overuse and neglect (Section 2). In each case, this has disrupted traditional markets, indigenous technologies and local communities.

5

Info Box 7. Effect of the European Community Agricultural Policy (CAP) of headage payments and price support (italic figures) in the livestock industry

The choices available to farmers can be represented as a system of balances. Each balance point represents a choice by the farmer in relation to the financial benefit. Traditionally, the choice of such things as the production system (arable/livestock), stocking rates and the output per head of stock was 'weighted' by local market prices and environmental conditions. CAP policy changed this relationship by introducing heavy 'weights' into the system of balances in the form of headage payments (payment per head of stock) and price support (payments per kilogram\litre of product). The way these weights relate to different sectors of the livestock industry are shown in the following diagram. The average net subsidies are indicated in the balance pans, with the components of these below the pans.

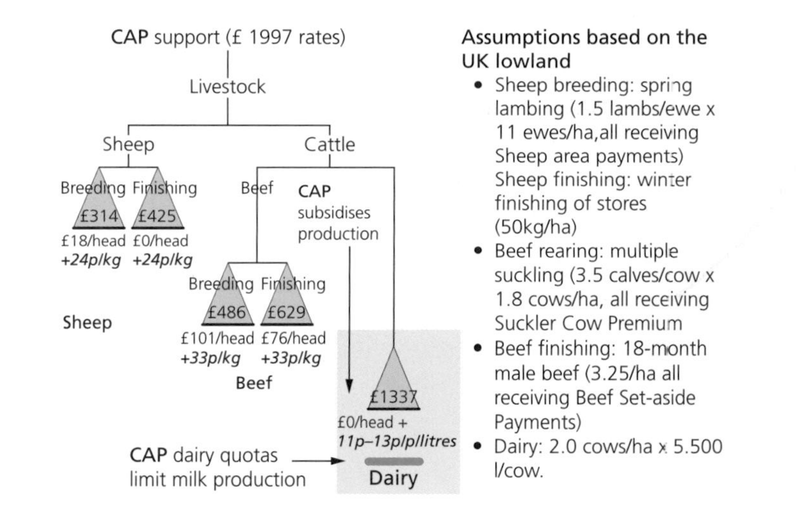

Allowing for geographical variation, the 'weighting' of payments tends to tip the balance towards the most intensive systems of production, that is: dairy (£1337) rather than beef (£629) and beef rather than sheep (£314). Headage payments tend to encourage farmers to keep more animals than they would otherwise. Price support encourages more production per animal and the use of larger breeds than would otherwise be the case. This externally applied CAP 'weighting' has resulted in a system of production that produces surpluses, over-rides natural environmental constraints and market conditions, and creates the most biologically impoverished pastures and the greatest pollution hazards in terms of slurry and silage effluent.

Attempts to reduce the intensifying effects of CAP support have used a system of quotas (stops) to limit production. These have addressed the effects of the payments system (restricting the swing of the scales) but not its intensifying nature (the cause of the problem). Essentially, the quotas can be represented by a stop on which a balance pan comes to rest. The consequence is that the system effectively becomes locked, and only amenable to radical shifts in policy. A way of addressing these causal issues is currently being evaluated by the European Commission and will be explained later in the Section.

In the case of overuse, the process of simplification resulted from the ability of price support to distort markets in such a way that production was neither limited by the market demand nor by the natural restrictions imposed by the land. The sequence of events seems to be that price support lifts market prices above free-market values, stimulating production, which helps to pay for the agri-chemicals and feed needed to override the landscape's natural constraints. This is why an indirect effect of the livestock subsidies referred to earlier include the 'improvement' of rough pasture, the use of fertilisers and the introduction of supplementary feed. Environmental degradation, however, (Section 2) suggest that, in the long run, intensification is not sustainable. This is because it pushes production into positions which strongly conflict with natural circumstances.

5.2 Attempts to control the adverse environmental effects of subsidies

Given the relationship between the current agricultural support mechanisms and the process of intensification, it is surprising that attempts to protect the environment (and incidentally to curtail agricultural expenditure in the form of subsidies), have addressed the effects of intensification rather its causes. Rather than tackling the cause of the problem (rewarding intensification), the various economic controls have merely tinkered with the problem by putting limits on production in the form of quotas and set-aside. Similarly, efforts to protect the environment have entailed using geographically isolated designations, such as SSSIs, and palliatives in the form of the agri-environmental measures, rather than tackling the economic forces which underlie the intensification of the agricultural system.

Some of the dynamics involved in current attempts to protect the environment from intensification are illustrated in Figure 29. This shows a simplified representation of the way incentives for both agricultural intensification and conservation interact with one another and with the environment. Two habitat curves are shown, a presumed 'original' distribution of Europe's habitats (dark curve), and an assumed 'current', distribution (light curve) which is largely created and maintained by CAP-funded

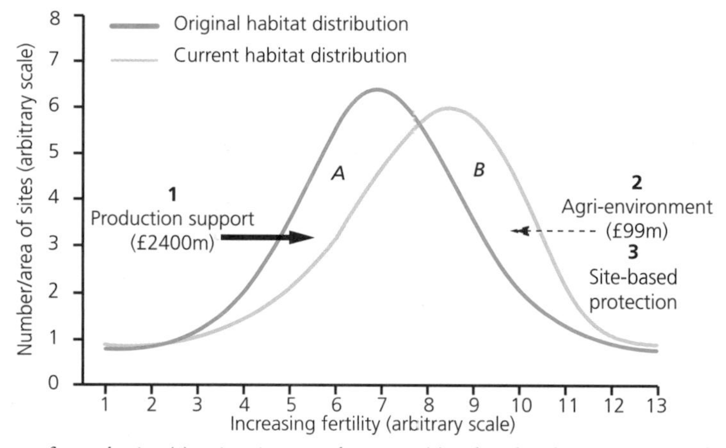

Figure 29. Pressures for production (1) agri-environmental measures (2) and site based conservation (3) (figures 1994/5), in relation to the distribution of habitats and species richness. Most species are adapted to coexist in the middle ranges of the 'original' distribution (mesic sites). Intensification has altered the 'original' site distribution, and has replaced mesic (species-rich) sites (A) with fertile (species-poor) sites (B)

intensive farming practices (bold arrow pointing to right). Working against this production-created shift in habitat distribution are a growing number of protective designations (Chapter 3) and a range of CAP funded conservation schemes (dotted arrow pointing to left).

The problems with the funding and protection arrangements set out in Figure 29 are threefold:

- Protective designations are largely ineffectual, in that they are a'representative' sample, have a limited coverage and are flawed as an approach to biological conservation (Chapter 3).
- Total environmental funding is considerably (2 orders of magnitude) less than the countervailing CAP expenditure on production.
- Even if expenditure on production and conservation under the CAP were more equally balanced, they would still be in opposition and thus both financially wasteful and self-defeating.

The imbalance and conflict between European support for production and conservation should not be surprising. The Common Agricultural Policy is designed for production not conservation, and its recently introduced environmental measures are simply 'bolt-on' features that do very little to correct its fundamentally production-orientated nature. The environmental measures so far introduced tend to deal with the consequences of environmental degradation rather than its causes. Because of this, they are often based on muddled thinking, fail to address key issues, and are sometimes ineffective. Some of these problems are illustrated by an assessment of the Test and Avon Valleys Environmentally Sensitive Areas (Info Box 8).

Info Box 8. Agri-environment schemes (UK): summary of performance of the Test and Avon Valleys Environmentally Sensitive Area schemes, Hampshire (1988-1997).

Schemes have made little difference to the biological quality of the valleys, possibly because:

- schemes have a poor rationale and confused aims;
- key actions (raising the water level) are beyond the scope of the ESA scheme;
- high commitment to established intensive practices, low income from ESA scheme and thus low take-up by farmers.

In the case of the Test valley, there are also:

- high-value fishing interests that diminish interest in potential ESA income, and;
- large areas are protected from change by SSSI status or National Trust ownership.

Source: Colin Tubbs (La Canada No 8, p 6-7, December 1997).

Table 15. Selected agri-environment schemes for the UK (1996 data). MAFF.

Funding Scheme	Eligible Area	Uptake (ha)		
Stewardship Schemes	unrestricted	106,806		(0.6)
Habitat Scheme	unrestricted	5,100		(0.03)
Environmentally Sensitive Area (No. 22)	1,149,208	80,292	[7.0]	(0.5)
Organic Aid Scheme	unrestricted	4073		(0.02)
Nitrate Sensitive Areas (No.32)	35,000	19,611	[56]	(0.1)
	Totals	215.882		(1.35)

Notes: figures in square brackets show uptake as a percentages of the eligible area; those on round brackets show uptake as a percentage of 'Total area on agricultural holding' (17,164,200 ha).

5

Table 16. Expenditure on agri-environment and other measures (UK) where the primary aim is the protection of the rural environment

	£ millions	% of total agri-support
1992-93	73.6	3.6
1993-94	93.8	3.2
1994-95	99.2	3.9
1995-96	117.5	4.1
1996-97	127.2	2.9

Hansard (Commons) written answer 24/01/1998 (pt 15). The lower % in 1996-97 follows BSE-related increases in the total budget

The European Union's Agenda 2000 proposals (July 1997) include plans to: reduce direct support to farmers and bring prices closer to those of the world market, and to develop a coherent and sustainable rural development policy. They also include plans to reinforce the existing agri-environment measures and to integrate them into support for production.

There are many kinds of CAP agri-environmental schemes throughout Europe, each dealing with different problems created by CAP policies that support intensive production. In the UK, uptake of the five most important schemes (Table 15) has been low - in total affecting less than two percent of the nation's farmland. Although the position looks more encouraging for stewardship (7%) and even good for the nitrate sensitive areas (56%), it should be remembered that these particular schemes only involve a small area of farmland.

Although the agri-environment initiatives are well-meant attempts to retrieve a worrying situation - and do show that something is being done - they are poorly funded (Table 16) and have to oppose powerful incentives for intensification (Figure 29). For these reasons, they are unlikely to have much impact on the farmland of Europe - except perhaps in those marginal areas where full-blooded intensification has never really been an option.

Fortunately, the cost of sustaining intensive production (Section 2) and our inability to protect the environment from its damaging effects (Section 3), are forcing us to rediscover that the interests of biological conservation, environmental protection, and sustainable production are probably best served by the simple but radical expedient of farming within the natural restraints of the land. This means farming in a way that reflects the natural potential of Europe's landscape resources. For the first time, the validity of this approach has been recognised in elements of the recent Agenda 2000 proposals.

5.3 Need for an integrated approach to conservation and land management

Taking the broad picture set out in Figure 29, it would seem that the way to reverse environmental degradation and achieve a sustainable environment, would be to:

- Reduce the level of support for intensive production, and make what remains environmentally neutral.
- Establish or re-equip modern systems of sustainable land-management based on long-established models, using the resources of the agri-environmental schemes, together with funds currently used for intervention and storage, and the support of the EC Structural Fund (objectives 1,5b and Leader II).
- Place less reliance on site-based designations, except as part of a wider strategy and as an interim means of identifying important resources.

Reforms along these lines could provide a number of environmental benefits. The agri-environmental measures for instance, would be more effective if the opposing incentives to intensify were less attractive (reduced support) and were less environmentally damaging (environmentally neutral). Moreover, if the money saved by cutting back on intensification could be used to re-equip modern versions of extensive farming, European agriculture would move a long way down the road to harmonising the interests of food production and biodiversity. Ultimately, this will lead to a more sustainable form of farming and a more credible form of conservation.

5.4 Reduction of support for intensive production

Although some of the habitat displacement shown by Figure 29 is attributable to social and technological factors, it is likely that much of it has been encouraged either directly or indirectly by post war land-use policies. In this context, the continuing reforms to the Common Agricultural Policy may begin to temper some of the worst effects of intensification. The first round of reforms (1992) de-coupled income support from market policy so that farm incomes could be maintained while price support was reduced for a range of commodities. The income support helped to maintain farming activity (especially in the less favoured areas), while the reduction of price support brought the price of EU produce more into line with World market prices.

The World Trade Organisation was established to administer and monitor world trade agreements. It has 132 member countries, is a form for trade negotiations and disputes, and provides training for developing countries.

It is likely that CAP reform will continue to bring farm prices closer to world market levels, in compliance with the requirements of the World Trade Organisation. Experience in the USA and New Zealand suggests that market alignment can have the effect of pushing farmers into further industrialising agriculture, which is a danger for Europe. On the other hand, market alignment will expand marketing opportunities for quality (locally distinctive) European produce.

More significantly, in the long term, it is possible that market alignment will have the effect of creating a closer relationship between production levels and the carrying capacity of the landscape. This change could be achieved by the simple expedient of making inputs of agrichemicals and feed increasingly unattractive financially. Under these circumstances, the plants and animals selected for intensive production may not be the most effective means of exploiting the lower fertility levels in the system. This may provide more scope (or even a greater need) for hardy, locally-adapted breeds of stock and crop - particularly on the less fertile soils, and the effect of this would be to diversify production, processing, markets, consumer choice and ultimately, the landscapes of Europe.

5.5 Environmentally neutral support for agriculture

Since production support is unlikely to be removed entirely, at least in the short-term, efforts will need to be made to make it as environmentally neutral as possible. This means providing support that avoids the intensifying effects already illustrated by livestock subsidies. An approach which addresses the problem of making support more sensitive to the environment is currently being examined in a study that focuses on livestock systems. The study is also looking at practical ways of integrating agricultural support and environmental management. The intention is to make existing support as production neutral as possible, and integrate it with a scheme of locally-appropriate environmental funding. The general approach being taken by the study is shown by Figure 30.

Figure 30. Options for the integration of environmental concerns into the CAP system of livestock support. Part of a study carried out for DGX1 of the European Commission by C.E.A.S. Consultants and the European Forum on Nature Conservation and Pastoralism. Approach was developed in six sites across Europe.

A European zones of agriculturally and environmentally similar areas

These will be based on the bio-geographical regions of the EU Habitats Directive further subdivided between lowlands and mountains. They will set a landscape-scale basis for defin ng the environment measures.

B Basic support payments

Tier 1 unconstrained production

Flat-rate area payment per AFH independant of type and number of stock, conditional only on producing the specified forage crop (rough grazing, permanent grassland etc)

Note 1:
(AFH)or Adjusted Forage Hectares, are area payment adjusted according to relative productivity of the land (in broad categories) ensuring that, in the change from headage payments and price support to area payments, there are no big winners or losers.

Note 2:
unlike headage payments (payment per head) and payment per kilogram (price support) area payments do not encourage overstocking or intensification and are thus more production-neutral

C Environmental payments

Tier 2 broad-brush environmental conditions

Tier 1, but with: for example, max/min stocking rates per AFH and max/min percentage of cattle within the livestock units (LU) (appropriate conditions for each zone)

Tier 3 Environmental management

Tier 2, but with: positive management (within the local agri-environment arrangements) for such things as seasonal grazing and the recreation of features: (appropriate conditions for each biographical zone)

At its heart, the study is motivated and guided by a number of simple ideas. These are:

- Environmental objectives should be defined at a landscape scale on the basis of similar agricultural and environmental conditions (A).
- Incentive for farmers to overstock would be removed if the headage and price support payments were to be replaced by a system of area payments. Faced with an area payment, farmers would tend to adjust the number and type of stock to suit the natural capacity of the land, thereby potentially reverting to a more sustainable, environmental-friendly form of land-use (B).
- Additional environmental objectives can be achieved (where appropriate) through a graduated (tiered) system of 'opt-in' incentives (C), paid in addition to the basic area payment.
- In intensive systems, farmers would have the option of farming to the market (unsupported) or receiving area payments with basic conditions attached.
- Other incentives should not conflict with those for conservation

Production-neutral support involves encouraging farmers to work within the carrying capacity of their land. In the case of beef production, this means gearing payments to land area (production-neutral) rather than numbers of cattle (production positive).

5

5.6 Re-equipping a sustainable agriculture

Reducing support for intensive production and introducing area payments may be sufficient in themselves to encourage a wide range of rural diversification. However, the changes that have affected the managed landscape of Europe over the last few decades have also altered the socio-economic make-up of the countryside - as indeed have other factors. In many areas, this may mean that sustainable systems of production will need to be restored and re-equipped, along with their associated processing and support industries, markets, and other key elements of the rural infrastructure.

Future options - the Cork (IRE) conference (November 1996) The Cork conference sketched out options for a European Rural Development Policy which were taken up in a paper on rural development published by the European Commission (DGVI) in November 1997

Resources for this 'enabling' process could come from a number of European sources. The European Union Structural Funds, for example, already have a growing financial commitment to rural enterprise under objective 1, 5b and Leader II (Info Box 9). In England alone the budget available for 1994/ 1999 under objective 5b is £410 million, which is £66 million more than it was for 1988-1993. There is also a possibility that money currently used for intervention and storage might also be used to stimulate rural development. As farm prices decline, these funds, together with private sector investment, could provide the necessary impetus to regenerate the rural economy.

This process of reform, however, should not tie farmers in the way that the CAP budget has done over the last few decades. On the contrary, within the natural constraints of the land, they need to be encouraged to respond to the market creatively and with confidence. For this reason, it is important that rural regeneration should be shaped by sustainable, locally-based initiatives which reflect the character of the landscape. This is a situation where governmental controls need a light touch. Their emphasis should be to provide a framework for the process of rural regeneration rather than define its course. Essentially, this is the message of the 1996 Cork conference, which set out some ideas for the reform of Community policy on rural development:

Info Box 9. European Community programmes under Objective 1, 5(a), 5(b), 6 and Leader II.

These help the development of viable rural communities and are financed by the European Structural Funds which attempt to even out disparities between richer and poorer areas.
Objective 1: deals with regions which lag behind the rest of Europe (<75% average EU GNP) and is concerned with diversifying farming and extending rural economic activity. This includes such things as restructuring to prepare for the free-market (Germany) and modernising the production of traditional products (Italy).

Objective 5(a): is concerned with adapting agricultural structures throughout the EU.

Objective 5(b): maintains and develops viable activities in areas which are having difficulties but are not covered by Objective 1.

The budget provision for the EU 12 countries under Objectives 1, 5(a), 5(b) and 6 have increased from 60.0 billion Ecu (1988-1993) to 141.1 billion Ecu (1994-1999), with a further 4.747 (1994-1999) billion Ecu for the three new member states (Austria, Finland, Sweden).
Leader II: succeeds Leader I, and supports rural development projects that are designed and managed by local partners. It can operate within 5(b) areas.
The budget provision for Leader II has increased from 0.4 billion Ecu (1988-1993) to 1.7 billion Ecu (1994-1999).

5

- Sustainable rural development will have a high priority in Europe, and will aim to reverse rural depopulation, stimulate employment, and respond to demands for better quality, and safer, food.
- Rural policy must be multi-disciplinary in approach and multi-electoral in its application, and must be decentralised, and based on the partnership approach.
- Support for economic and social diversification will provide a framework for self-sustaining private and community-based initiatives.
- Rural development policy must be transparent, coherent, involve simple procedures, and must take account of a region's special characteristics.

Despite these aspirations, there is still a real possibility that rural development initiatives will tie farmers down to inappropriate management activities much in the same way that production-orientated agricultural policies have done for the last few decades. This danger can be highlighted by the different ways that Scotland and Portugal apply the same EU agri-environmental regulations to their respective zonal programmes (Table 17).

The Portuguese approach to the Castro Verde zonal plan very much reflects the spirit of the Cork Agenda. It is based on the partnership approach and is locally controlled by farmers. It also has a devolved, uncomplicated administration which responds to local needs and can adapt quickly; moreover, its 'bottom-up' approach is based on a traditional system of land-management that is inherently sustainable. It has therefore become an effective form of rural development which enjoys the enthusiastic support of the local farming community. The Argyll Islands ESA, by contrast, is based

Table 17. Contrasting approaches to Zonal programmes in western Scotland and southern Portugal using the Agri-environmental Regulation 2078/92.

	Argyll Islands ESA	Castro Verde Zonal Plan
Establishment of aims and ground rules	**Centralised:** Scottish Office (Agriculture, Environment and Fisheries Department	**Diversified:** Committee comprising the Ministries of Agriculture and Farmers' Association, and two Non Government Organisations
Control and Administration	**Centralised** (top-down) Scottish Office (Agriculture, Department)	**Localised** (bottom up): Farmers Association in consultation with the local advisory committee
Basis of Control	**Prescriptive** (and complex): background rules (20, A4 pages) and a detailed management plan.	**Responsive** (and simple): background rules (3, A4 pages) and an agreement to farm using established traditional methods
Monitoring and Review	**Biennial:** review carried out by the Agriculture, Environment and Fisheries Department of the Scottish Office, in consultation with other departments	**Annual:** Farmer deals directly with the local support officer (employed by the Farmers Association)

Source: G Jones (La Canada No8 December 1997, p4&5)

on a 'top-down', centralised approach to rural development. It does not involve farmers in the planning or administration of the project, nor does it look to the needs of the underlying land-management system. Instead, it attempts to 'preserve' habitats by way of complicated prescriptions and management plans, rather than by supporting the traditional forms of land-management that produced the habitats in the first place.

The problem with the Argyll Islands approach, is that we simply do not know enough about ecological relationships to be able to produce a realistic prescription for even one habitat, let alone for a mosaic of interacting habitats - the same is also true for human communities and cultural systems. In order to cope with this complexity, rural development initiatives need to step back from their preoccupation with the detail and consider the land-management systems that produce the effects we value. Essentially, this is how the Castro Verde zonal plan operates. The scheme was introduced to protect one of Portugal's largest populations of steppe birds. These depend upon an extensive pseudo-steppe which is the product of centuries of low-intensity cereal growing. The plan, therefore, supports traditional farming practices as a natural, un-fussy way of conserving the 'pseudo-steppe'. Using the agri-environment regulations in this way enables Portugal to make the link between 'environmentally-friendly' agriculture and premium products such as hams, high quality olive oil and cheeses.

In rural development and conservation alike, it seems that once the central importance of traditional systems of land-management is accepted, many hitherto intractable problems resolve themselves. For example, not only does it become possible to see ways of integrating the best interests of quality food production and conservation, it becomes clear that the two are mutually dependent. It also becomes easier to see that the only practical way to secure a sustainable approach to land-management is to remove the legislative and economic constraints that prevent farmers from farming efficiently within the 'natural' limitations of the land.

Summary:
Attempts to counter the damaging social, economic and environmental effects of intensive production have largely failed because they have a narrow, symptomatic approach that does not address causal issues. Together with the growing economic burden of intensive production, these failures have forced a comprehensive, European-wide review of land management policy, initially in the Agenda 2000 reforms. This ongoing process appears to embrace a systems approach that looks at causal factors.

The review is currently examining ways to: remove incentives to over-production; integrate environmental protection into agricultural support; and to create a coherent framework for rural regeneration.

Providing the resulting measures are ecologically sound, implemented sensitivity and funded appropriately, it should become progressively easier to farm efficiently and sustainably; that is, within environmental limits, using locally-adapted crops, stock and technologies.

Since this approach to land management is the basis of traditional farming, it should not only remove the conflict between production and conservation, but also promote landscape, cultural and biological diversity.

6 Abbreviations used in this publication

BSE	Bovine Spongiform Encephalopathy ("Mad Cow Disease").
CAP	Common Agricultural Policy (of the European Union).
CBD	Convention on Biological Diversity.
EC	European Commission (i.e. "Civil Service" of the European Union).
EFNCP	European Forum on Nature Conservation and Pastoralism.
EU	European Union.
FAO	Food and Agriculture Organisation (United Nations).
GMC	Genetically Modified Crops.
IUCN	International Union for the Conservation of Nature - The World Conservation Union.
MAFF	Ministry of Agriculture and Food (UK).
NVC	National Vegetation Classification (UK).
SSSI	Site of Special Scientific Interest (UK).
UK	United Kingdom.
UN	United Nations.
UNESCO	United Nations Education, Scientific and Cultural Organisation.
WTO	World Trade Organisation.

7 Further Reading

Agenda 2000 and Prospects for the Environment (1998).
Report of the seminar organised by the European Forum on Nature Conservation and Pastoralism at COPA, Brussels on 3rd February 1998. Editors, Steve Goss, Eric Bignal & Mike Pienkowski (eds). EFNCP Occasional Publication 16.

European Environmental Almanac (1995).
Principal Editor Jonathan Hewet, Earthscan, London.

Farming and Birds in Europe (1997).
The Common Agricultural Policy and its implications for bird conservation. Edited by Deborah J. Pain and Michael W. Pienkowski. Academic Press, London.

La Canada: Newsletter of the European Forum on Nature Conservation and Pastoralism.
Editors, Eric Bignal and Andrew Branson, ISSN 1027-2070. Subscription details EFNCP, Rooks Farm, Rotherwick, Hampshire RG27 9BG.

Sustainable Agriculture in the UK (1996).
A report written for the CPRE and WWF-UK by a team from the Institute of Environmental Policy and Adrian Phillips. Available: WWF--UK, Panda House, Weyside Park, Godalming, GU7 1XR.

The Common Agricultural Policy and Environmental Practices (1996).
Proceedings of the seminar organised by the European Forum on Nature Conservation and Pastoralism at COPA, Brussels on January 29th 1996. Editor, Karen Mitchell. EFNCP Occasional Publication I.

The Commoners' New Forest (1944).
Kenchington, F. E. Hutchinson & Co Ltd, London.

Europe's Environment: The Dobris Assessment (1994).
A comprehensive review of Europe's environment, published by the European Environmental Agency, Copenhagen 1994.

The Nature of Farming (1994).
Low intensity Farming Systems in Nine European Countries: Beaufoy, G. Baldock, D. Clark, J. Published by the Institute for European Environmental Policy, London.

8 Useful Web Sites (verified: 17th November 1998)

UN. FAOSTAT http://apps.fao.org/
The on-line and multilingual database of the United Nations Food and Agriculture Organisation (FAO). It currently contains over 1 million time-series records covering international statistics in the following areas: Production, Trade, Food Balance Sheets, Food Aid Shipments, Fertiliser and Pesticides, Land Use and Irrigation, Forest Products, Fishery Products, Population, Agricultural Machinery. Note: some data are 'unofficial' or based on 'FAO estimated'and are subject to minor changes as data quality and reliability improve; see relavent FAOSTAT notes for details.

EC. DG VI
http://europa.eu.int/en/comm/dg06/index.htm
Home page of the Agricultural Directorate of the European Commission. Information on policies, budgets and news. EC. DG XI (Environment)

EC. DG VI
http://europa.eu.int/comn/dg11/index_en.htm
Home page of the Environmental Directorate of the European Commission. Information on policies, budgets and news.

EC. DG XI
http://www.biodiv.org/convtext/cbd0000.htm
Text of the Convention on Biological Diversity (1992).

MAFF http://www.maff.gov.uk/
Home page for the Ministry of Agriculture, Fisheries and Food (UK) information and statistics on Farming, Environment and BSE.

EFNCP http://www.efncp.org
The European Forum on Nature Conservation and Pastoralism brings together ecologists, nature conservationists, farmers and policy makers. This non-profit network exists to increase understanding of the high nature conservation and cultural value of certain farming systems and to promote their maintenance.

EEA http://www.eea.dk/
The European Environment Agency - an important source of information on Europe's environment.

9 Acknowledgements

The authors are grateful for ideas, material and editorial guidance provided by Eric Bignal and Steve Goss (the other speakers at the John Mason conference), and also for the support of other members of the European Forum on Nature Conservation and Pastoralism network, particularly Gwyn Jones, Jenni Tubbs, Peter Eden, Rainer Luick and Gun Rudquist.

Addresses of Contributors

Dr C. Hindmarch, European Forum on Nature Conservation and Pastoralism, 66 Hunters Crescent, Romsey, Hampshire, SO51 7UJ, UK [e-mail c.hindmarch@btclick.com].

Dr M.W. Pienkowski. Director, European Forum on Nature Conservation and Pastoralism, 102 Broadway, Peterborough, PE1 4DG, UK [e-mail pienkowski@cix.co.uk].